A Brief History of Bacteria

Of Bacteria

The Everlasting Game Between Humans and Bacteria

A Brief History of Bacteria

The Everlasting Game Between Humans and Bacteria

Written by
Daijie Chen and **Xiuping Qian**
School of Pharmacy, Shanghai Jiaotong University, China

Translated from the Chinese by
Qingli Hu and **Youjia Hu**

Illustrated by
Yuankai Xue

Image editing by
Bing Ni and **Yu Yin**

Chemical Industry Press

World Scientific

Published by

World Scientific Publishing Co. Pte. Ltd.
5 Toh Tuck Link, Singapore 596224
USA office: 27 Warren Street, Suite 401-402, Hackensack, NJ 07601
UK office: 57 Shelton Street, Covent Garden, London WC2H 9HE

and

Chemical Industry Press
No. 13, Qingnianhu South Street
Dongcheng District, Beijing 100011
P. R. China

Library of Congress Cataloging-in-Publication Data
Names: Chen, Daijie, 1957–   author. | Qian, Xiuping, 1964–   author.
Title: A brief history of bacteria : the everlasting game between humans and bacteria /
    Daijie Chen, Xiuping Qian.
Other titles: Xi jun jian shi. English
Description: New Jersey : World Scientific, 2017.
Identifiers: LCCN 2017037692 | ISBN 9789813225152 (hardcover : alk. paper)
Subjects: | MESH: Bacteriology--history | Bacteria--pathogenicity
Classification: LCC QR74.8 | NLM QW 11.1 | DDC 616.9/201--dc23
LC record available at https://lccn.loc.gov/2017037692

British Library Cataloguing-in-Publication Data
A catalogue record for this book is available from the British Library.

For any available supplementary material, please visit
http://www.worldscientific.com/worldscibooks/10.1142/10573#t=suppl

Typeset by Stallion Press
Email: enquiries@stallionpress.com

Foreword

Bacteria are a kind of microorganism, short in shape, and simple in structure. It is the most widespread and populous life form in nature. Bacteria are both beneficial and dangerous to humans, animals, and the environment. There are "superbugs" which are highly resistant to antibiotics and can cause life threatening diseases. There are also faithful friends like Bifidobacterium which helps to maintain the balance of intestinal flora and resist pathogens. To reveal the truth about bacteria in an objective and lively manner and promote public understanding of this form of microorganism, such is the contribution of Chinese popular scientist Gao Shiqi. He has drawn much public attention to bacteria through a series of popular science publications such as *The Everyday of Bacteria* and *An Adventure into the World of Bacteria.*

Professor Daijie Chen has dedicated over 30 years of his life to the research, development, and education of medical microbiology. He is very knowledgeable on the subject of bacteria resistance and has much experience in the development of antibiotics. He is also an active writer. *A Brief History of Bacteria* tells a story of how bacteria, the enemy of humans, cause disease, pain, and even death; it also tells how bacteria are also our friends that benefit our lives. This book is rich with simple language and lively illustrations. Readers will find it very informative.

There have not been many worthy popular scientific books recently in China. This book by Professor Chen is a successful attempt in the area. It will probably be an efficient tool to promote knowledge on bacteria. Thus, I am happy to write its foreword.

Han Qide
President of China Association of Science and Technology

Preface

As a scientist, I always believe that apart from my research, it is also my responsibility to convey my knowledge and thoughts to lay readers with light language, so as to increase the bond between scientists and the public. Science should step out of ivory towers and libraries, and enter the heart of those who crave it.

Based on my long-term research interest, I have always wanted to write a popular science book about bacteria and medicine. Thirteen years ago, East China University of Science and Technology published my humble work *Antibiotics and Resistance of Bacteria*. The content of the book is very professional. But from a popular science point of view, this is a book about the never-ending war between humans and bacteria. I planned to write this book in the same theme. But as I was developing and improving my ideas, I increasingly felt that this would not be a fair approach to bacteria. After all, apart from being the enemy that threatens human lives, bacteria are also the friend of human civilization and have made significant contributions to human progress. Keeping the war between bacteria and human as the main story, while also mentioning their contributions—this is a fairer attitude toward those tiny fairies. It will also open a door to the audience through which they can comprehend the rights and wrongs of bacteria.

To realize this long-cherished wish, I have been collecting popular scientific books on life science, pharmacy, and medicine for more than a decade. On the one hand, I would like to see if my idea "clashes" with any of theirs and whether it is worth publishing; on the other hand, I was exploring the writing techniques of these popular science titles so that I can objectively evaluate my own ability and decide if I am capable of this not-so-easy challenge. It is my belief that a good piece of popular science work should be based on the author's deep comprehension of these sophisticated theories, complicated experiments, and rigorous arguments. Furthermore, the author should convey the scientific facts and rules with attractive language, to ensure that readers will not only understand easily, but also remember by heart, and voluntarily communicate their knowledge to others. Such an effect is hard to achieve.

"Bacteria" is a familiar term to all households. However, my dear readers, do ask yourself: how much do you know about them? Perhaps you will associate the word bacteria with disease and are horrified by its ferocity. Indeed, the main topic of this book is how bacteria, the enemy of humans, cause disease, pain, and even death. It tells you about the eternal war between humans and bacteria. You will be surprised to discover that pathogenic bacteria have very refined battle strategies and equipment to attack human bodies; that human immune system bravely launch prompt and effective defence against invasion inside the human body; that the discovery and application of "magic medicines" penicillin and streptomycin are so brilliant; or that the scientists have designed their medicines according to the enemy's abilities. Yet bacteria are cunning. To avoid being persecuted by medicine, they launch resistance with abundance, which is both exciting and thrilling. To win this war, humans have been fighting superbugs that are invulnerable and threatening. To whom will victory belong? Dear readers, maybe you do not know that bacteria can also be human's critical friends. Without bacteria there will be no human lives, nor vast social developments, nor the colorful world we live in today. Therefore, this book also tells how bacteria are beneficial to humans. Readers will also see that some bacteria live in harmony with the human body and are indispensable to our health; that they also help in refining biological energy in the post-fossil era and in producing tasty fermented food; that they provide biofertilizer for modern agriculture and participate in biodegradation and biogovernance to protect and restore the environment. All these are significant achievements of bacteria.

Dear readers, after you finish reading this book, you will know how bacteria are both lovable and detestable. You will see how much intelligence and bravery it requires for humans to discover, use, and conquer bacteria. You will then be able to judge the merits and demerits in the war between bacteria and humans, and decide who will be the final victor. Meanwhile, after you finish reading, you will surely be more rational and knowledgeable as to how to cooperate with the doctor when bacteria invades your body. Hopefully, you will combine this knowledge with your personal experience and share them with your family and friends, so that there will be more wisdom among us. I believe this is a worthy book for you to keep at hand. I have received much support while writing this little book, and I have adopted many valuable recommendations. In the early stages, some 2003 graduates majoring in microbial and biochemical pharmacy from Shanghai Jiao Tong University and Shanghai Institute of Pharmaceutical Industry helped with the literature review and with writing certain parts of this book; later, Professor Qian Xiuping from Shanghai Jiao Tong University joined us and made this book more logical, systematic, and readable; afterwards, with the cooperation of amateur artist Xue Yuankai, who has a background in biology, the text was converted into lively pictures. Professor Ni Bing from East China Normal University and Mr. Yu Yin from Shanghai Laiyi Research Centre for Biochemical Pharmacy both helped to make the *scanning electron microscope* (SEM) photographs and light micrographs. In addition, this book cited some figures from *Introduction to Bioindustry*, edited by Dr. Mei Ge and I. Some other pictures are provided by friends. I would like to express my sincere thanks here. Staff at the Chemical Industry Press have contributed valuable advice for editing and improvement based on their extensive knowledge and publishing experience. Their passion has been an encouragement for me to complete this book.

The author acknowledges the limitations of his knowledge. Suggestions from readers are welcome, should there be anything improper or incorrect in this book.

Daijie Chen

Contents

Foreword   v

Preface   vii

Chapter One: Introducing Bacteria   1

Chapter Two: Friend or Foe?   39

Chapter Three: A Silent Battle in the Body   117

Chapter Four: A History of the Hard and Difficult
War Against Bacteria   145

Chapter Five: A Protracted Tug of War   215

Chapter Six: Who will be the Winner? The War
Continues   247

Index   279

Chapter One

Introducing Bacteria

In nature, there are animals and plants of all forms and colors. But do you know there is also a mysterious group of tiny individuals, too small to be seen with the naked eye? These are called microorganisms. Microorganisms are tiny creatures that are less than 0.1 mm in size. There are many types of microorganisms. What people hear, see, and get in contact with are virus, mold, yeast, and bacteria. In this chapter, you are introduced to one member of the microorganisms' world — bacteria.

Bacteria are the earliest "residents" on Earth. Traces of them date back 3.5 billion years, whereas human beings have a history of only several millions of years. The tiny bacteria possess five characteristics that are impossible among higher organisms: (1) they have small volume but large surface area; (2) they absorb considerable nutrition and transform it quickly; (3) they are full of vitality, reproducing at high speed; (4) they constantly vary and are good at adapting; and (5) they have a great variety distributed in many places. They are everywhere. With proper methods, people can find these tiny fairies in almost all corners of the world. Figure 1-1 is a personification of bacteria sojourned in animals, plants, soil, water, and air, which can only be observed through microscopes. These bacteria include rumen bacteria in cows, *Bifidobacterium* in children's stomachs, cyanobacteria in oceans, enterobacter in birds' stomachs, sheath blight bacteria in ailed rice, and the oil-degraded microorganisms that consume petroleum.

Figure 1-1: Tiny fairies everywhere.

What do Bacteria Look Like?

Bacteria are single-celled organisms, namely one cell is one organism.

We could hardly see how bacteria look like with our eyes. But with the help of microscopes, we can see clearly their various shapes and forms. They most commonly appear in spherical, rodlike, or helical shapes and are, respectively, called coccus, bacillus, and spirochete.

Coccus

In case of bacterial cell division, when the new organism exists independently, it becomes single coccus, such as *Micrococcus luteus*. When two cells are arranged in pairs, they are called diplococcus, for example *Pneumococcus*. When multiple cells form a chain structure, they

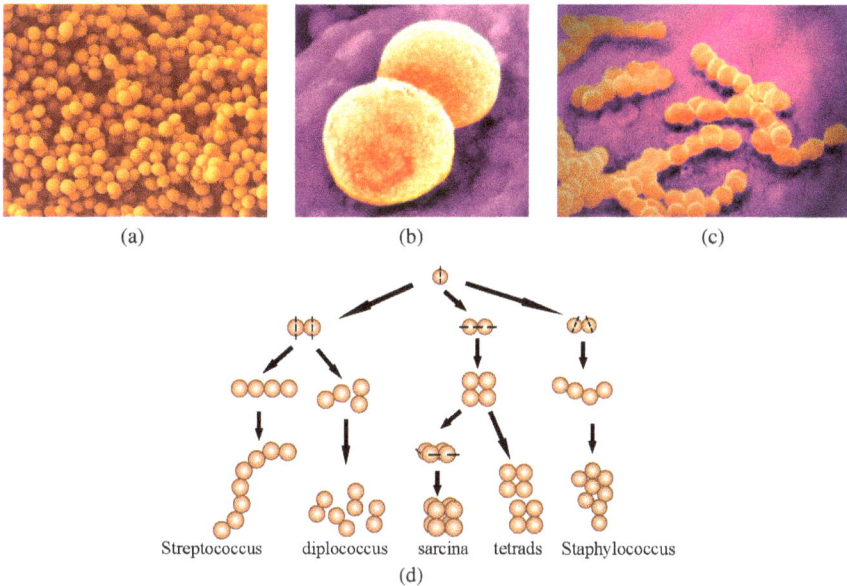

(a)   (b)   (c)

Streptococcus   diplococcus   sarcina   tetrads   Staphylococcus

(d)

Figure 1-2: Coccus and its reproduction and order. (a) *Staphylococcus aureus* (electron microscope; (b) *Pneumococcus* (computer simulated); (c) *Streptococcus lactis* (computer simulated); and (d) reproduction and order of coccus.

are called streptococcus, for example *Streptococcus lactis*. When divided twice, four cells link into "⊞" shape and become tetrads, for example *Micrococcus tetragenus*. When the cell divides in three vertical directions into eight, the split cells pile together in the shape of a Rubik's cube, which is called sarcina, for example *Sarcina lutea*. When there are no specific directions of division, the new organisms appear in grapelike clusters called staphylococcus, for example *Staphylococcus aureus*. Figure 1-2(a) is a scanning electron microscope (SEM) photograph of *Staphylococcus aureus*; Fig. 1-2(b) and (c) are computer-generated figures of *Pneumococcus* and *Streptococcus lactis*. Figure 1-2(d) shows the division and arrangement of coccus.

Bacillus

The shape of the two ends of bacilli varies. Some are round and blunt, some are flat, and some have a sharp tip. The length-to-

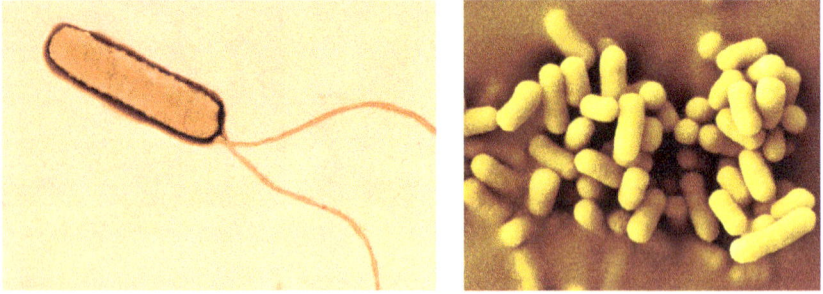

Figure 1-3: *Escherichia coli* (left, scanning electron microscopy; right, transmission electron microscopy).

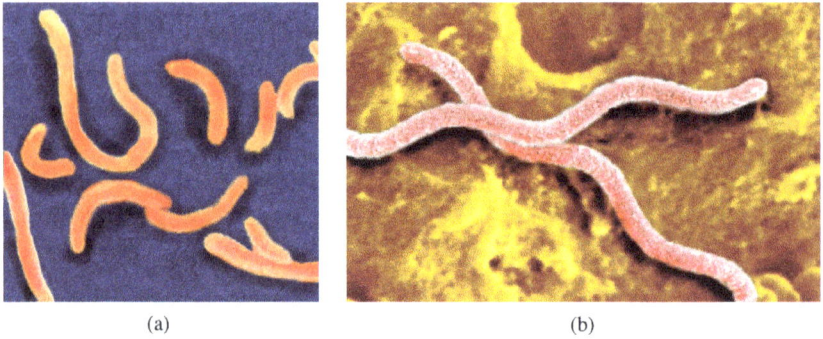

(a)   (b)

Figure 1-4: Helicobacter: (a) comma bacillus and (b) *Helicobacter pylori.*

diameter ratio among bacilli also varies. Some appear short and chubby, some appear tall and slim, and some have beard-like flagellum and pilum. Figure 1-3 is an SEM photograph of *Escherichia coli.*

Helicobacter

Helicobacter are the "dancing bacillus", or bacterial cells with a curved shape. They are divided into two types according to the degree and hardness of their curves. The first type is vibrio with short cells and a single, archlike curve, for example *Vibrio cholerae.* The second type is helicobacter. The cell curves more than twice, like a spiral, and is usually hard, for example *Helicobacter pylori.* Figure 1-4 shows computer-generated helicobacter.

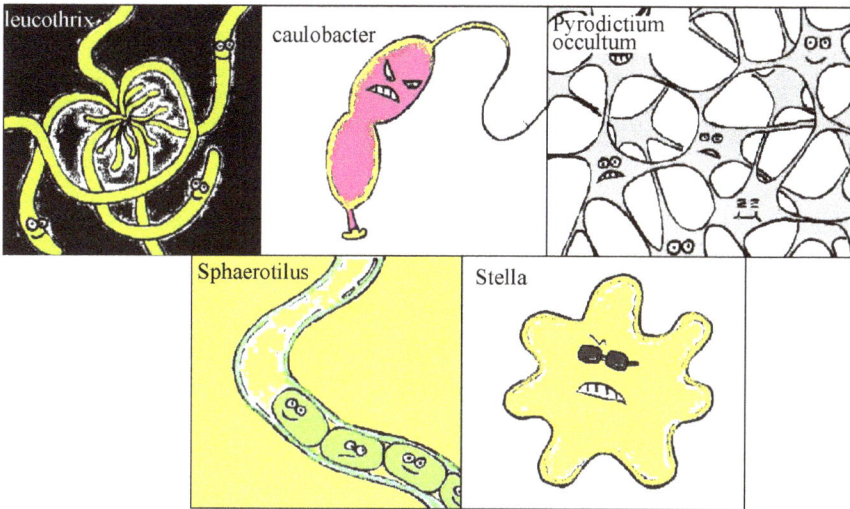

Figure 1-5: Bacteria with special shapes.

Apart from these, bacteria have many other shapes, for example, bacteria with appendage and handle, or bacteria shaped like thread, star, or rectangle. Figure 1-5 shows bacteria of special shapes.

In different phases and under different conditions, the shape of bacteria changes considerably. Generally, young bacteria or bacteria growing under suitable conditions appear in a certain neat and normal shape. Old bacteria or bacteria growing under unusual conditions will display irregular shapes. Compared to animals and plants, the changes in shape of bacteria are more difficult to comprehend.

How Big are Bacteria?

Bacteria are small and light. They are so small that you cannot see them with naked eyes; they are so light that you cannot weigh them.

Bacteria can only be seen when enlarged hundreds or thousands of times under a microscope. We can measure the size of a bacterium using micrometer under a microscope; or we can use projection or photograph, and measure the enlarged figure. The size of bacteria is measured in micrometers: 1000 micrometers is equivalent to 1 millimeter.

Take *Escherichia coli* for example. It has an average length of 2 μm and a width of 0.5 μm. 1500 *Escherichia coli* lying head to toe in a line is only as big as a sesame seed (3 mm); 120 *Escherichia coli* standing shoulder to shoulder is just as wide as one single hair (60 μm).

Biggest bacterium

The April 1999 issue of *Science* magazine reported the largest bacteria ever discovered in history. This coccus has a 0.1–0.3 mm diameter on average, and the largest can reach 0.75 mm. This is 100–300 times larger than normal coccus (Fig. 1-6). Normal bacteria, in comparison with this huge bacterium, appear like newborn mice in front of a blue whale. This bacterium was discovered by Schultz, a biologist at Max Planck Institute for Marine Microbiology in Germany, in the riverbed sediments in Namibian coast along Southwest Africa. This is the largest bacterium in the world discovered by far that is perceivable to the naked eye.

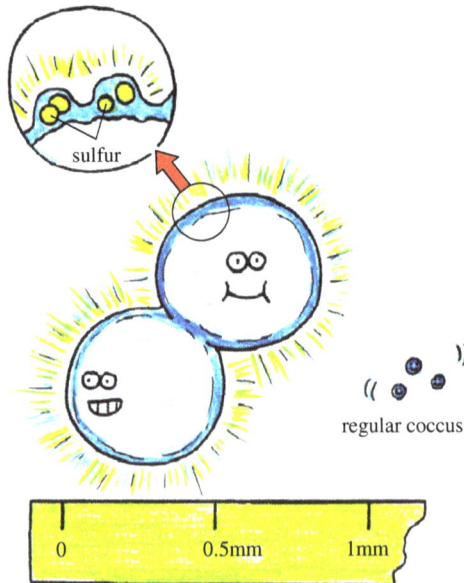

Figure 1-6:   Namibia sulfur pearl bacteria. Arrow indicates the reason for the shine in bacteria — considerable sulfur in bacteria.

Generally, this bacterium lives in seabed sediments with high concentration of hydrogen sulfide. It has a bright white color due to sulfur particles. When arranged in a roll, they look like a shining pearl necklace. Therefore, Shulz and other researchers named this bacterium *Thiomargarita namibiensis.* The sediment where this bacterium lives lacks oxygen but is rich in nutrition. There is also considerable hydrogen sulfide. Although hydrogen sulfide is poisonous to animals, it is food for this bacterium, because the latter can oxidize sulfur with the nitrate in its cell. The discovery of this bacterium has provided more precise evidence of the coupling effect between sulfur cycle and carbon cycle. These two main cycles in the ocean have been considered irrelevant up until recently.

There are also amazing microorganisms living under extreme conditions. For example, there are ancient bacteria living in springs with a temperature of more than 100 °C; there are bacteria in the Antarctica seabed and glacier, where temperature is below zero. The discovery of these microorganisms is exciting not only to those studying the origins of life but also to biologists.

Smallest bacterium

In 1997, Finnish scientist Kajander and his colleagues, while culturing mammalian cells, discovered a bacterium that could pass through the 100-µm bacteria filter. This bacterium is normally 1/20 the size of normal bacteria. The smallest bacterium discovered by far, it is accordingly named nanobacterium. Comprehensive research has revealed that nanobacterium is an amazing tiny thing that is capable of self-duplication and mineralization. It is also related to many diseases. Scholars from NASA believe it causes kidney stones in astronauts, as it exists at the center of calcium phosphate in the stones. To study the features of nanobacterium, NASA placed the bacterium in bioreactors that simulate outer space environment. Under microgravity, nanobacterium duplicates five times faster than under normal gravity. Nanobacterium can be infected among astronauts living in constrained space. This amazing little thing is also found in other diseases such as Alzheimer, heart disease, prostatitis, and some cancers.

The Little Fairy Everywhere

Although not perceivable with the naked eye, bacteria leave their traces everywhere on Earth, from the flamboyant city to tranquil forest, from snow-covered mountains to vast seas, from the warm and rainy tropics to uninhabited deserts, from freezing polar zones to boiling volcanoes, from the surface of animals and plants to their insides, from dust floating in the sky to fossils lying thousands of years underground. Bacteria have made their home everywhere.

Humans seem to live in a sea of bacteria. When you are playing or studying, there are numerous bacteria at your company; even in your tidy dorms, there are countless bacteria sharing the space with you. The environment we live in and the things we touch are often contaminated by bacteria, such as cleaning cloth, garbage bags, curtains, door handles, cutting boards, coins, public telephones, books and newspapers, tables and chairs, switches of electronic devices, bedroom furniture, and even some food. Research shows that a dirty hand carries 400,000 bacteria; among a random selection of 700 RMB notes, *Escherichia coli*, indicator of intestinal bacterial infection, are detected on 440 notes.

Bacteria in human bodies

Do not assume that bacteria will not like you, if you are a clean person; and do not assume that the less bacteria on your body the better. Scientific research shows that once you leave your mother's uterus, all types of bacteria attempt to "invade" each part of your body by various means, and they grow and regenerate, feeding on the nutrition in your body. Scientists speculate that on an adult body reside up to thousands of billions of bacteria, of more than 400 different types. The number of bacteria surpasses 90% of the total living cells of the human body (this includes all the living cells that make up the human body, as well as the microorganisms living inside the human body). Bacteria's favorite camping places include mouth, nostrils, intestines, and the skin.

Mouth

How many bacteria are there in the mouth? There are no bacteria in the mouth of a newborn. But as the baby announces his or her arrival into the world with a cry, bacteria, along with air, enter his or her mouth and start inhabiting there. Later, when feeding on milk, either through breast or feeding bottle, more and more bacteria enter the baby's mouth. Specialists estimate that in a clean mouth, 1000–100,000 bacteria reside on the surface of each tooth; in a not so clean mouth, there can be 100 million to 1 billion bacteria; 1 mg of plaque contains approximately hundreds of millions of bacteria.

Why there are so many bacteria in the mouth? It is because the mouth continuously secretes saliva. Saliva has 100% humidity and a temperature of 37 °C. Together with breathing, speaking, and eating meals, this is a greenhouse for bacteria, providing sufficient nutrition and oxygen for their growth and reproduction.

Like humans, bacteria have different races and groups, and they occupy their own living spaces in the mouth. Some live together at the cheeks; some gather at the ventral tongue; some prefer the back of the tongue; and anaerobic bacteria live in groups between the teeth.

Normally, despite staphylococci, streptococci, *Escherichia coli*, and other hundreds of types of bacteria residing in the mouth, the body can still be perfectly fine. Some bacteria are even beneficial by assisting food digestion. But when the body is exhausted and the immunity level is low, or when it is stimulated by other factors, bacteria will multiply in large numbers, creating a focus of chronic infection. The most common symptom is periodontitis. As the mouth is at the upstream of the respiratory and digestive tract, mouth infections can easily lead to sore throat, tonsillitis, tracheitis, pneumonia, ulcer, tuberculosis of the intestines, and many other diseases. Meanwhile, because of anatomic reasons, mouth infection can affect nostril and middle ear. Some researches show that *Helicobacter pylori*, which is responsible for gastritis and ulcer, can be detected not only in the pylorus but also in mouth, teeth, and saliva.

Nostril

Bacteria in the nostril are all trapped in nasal hair, which filter the air we breathe in. These bacteria are often the grapelike *Staphylococcus aureus*. Deep in the nostril, there are many chainlike streptococci. Meanwhile, under good health conditions, other bacteria detectable in the nostril include pseudodiphtheria. According to expert estimation, 500 bacteria are breathed out every minute, and 4000 each time when you sneeze.

Normally, the nostril can ensure that the number of those bacteria stays within acceptable levels to the human body. When encountered with triggering factors such as cold, rain, or exhaustion, the detoxification function of the nostril declines, and bacteria can stay in the nostril for longer time and multiply. This can easily lead to sinusitis. When sinusitis undergoes serious pathological changes, it may expand and invade neighboring organs and leads to osteomyelitis, orbital cellulitis, soft meningitis, brain cyst, or even septicemia.

Intestines

Among the thousands of billions of bacteria inhabiting human bodies, 95% live in the intestines, especially in the large intestine. To put it vividly, bacteria in the human body weigh up to 1 kg, almost 30%–50% of the weight of dry excrement. On the basis of their influence on the human body, these bacteria fall into three categories: the beneficial, the harmful, and the harmless.

Beneficial bacteria protect human body's nutrition absorption and health conditions. At 4–8 h after meal, 70% of the food digestion and absorption relies on intestinal bacteria. Meanwhile, some beneficial bacteria detain the growth of harmful bacteria and resist pathogens infection; they synthesize vitamin B, which is necessary to human bodies; they generate organic acids; they stimulate intestinal peristalsis, promoting defecation and avoiding constipation; they detain decaying in the intestines and purify intestinal environment; they decompose toxic and carcinogenic material and improve the immune system; and they also reduce the blood cholesterol level and slow down aging.

Bifidobacterium and *Lactobacillus acidophilus* are the two main beneficial bacteria for the human body. *Bifidobacterium* are obligate anaerobes. They live in the large intestine where oxygen is less. *Lactobacillus acidophilus* are facultative anaerobes. They live in the small intestine. Small amount of oxygen is not harmful to its growth.

Harmful bacteria in the intestines include *Clostridium perfringens*, Villanelle, Bacteroidetes, and *Pseudomonas aeruginosa*. They decompose food remains in the intestines and produce ammonia, amines, indole, skatole, nitrosamines, thiols, and other toxic compounds. These compounds, absorbed by the human body in a long term, will accelerate aging, reduce immunity, and cause various types of diseases. *Helicobacter pylori* in the stomachs can cause gastritis, which is the most common reason for ulcer. Along with other factors, it can also cause stomach cancer.

The amount of *Bifidobacterium* in human bodies differs by age. Babies have the largest amount, with bifidobacteria making up 95% of the total amount of bacteria. This decreases with age. With increasing age, the amount of bifidobacteria decreases and more harmful bacteria occupy the intestines. Ailing people have a similar intestinal environment with aging, where there are more harmful bacteria than beneficial ones. In general, most healthy people have more beneficial bacteria than harmful ones. Therefore, *Bifidobacterium* is a barometer for health conditions.

Generally, there is check and balance among the three types of bacteria. Once the balance is broken (e.g., severe illness, after operations, chemotherapy, radiotherapy, long-term use of antibiotics, aging, depression, and immunity problems), harmful bacteria take the upper hand. The intestines cannot function properly, and the body gets sick easily. *Escherichia coli* is a common bacteria that normally live in peace with the human body and synthesize the necessary vitamin B. But when they multiply too fast, they can also cause harm by producing toxic chemicals that lead to flatulence, diarrhea, and other diseases. According to medical research, increasing percentage of beneficial bacteria in human bodies and detaining excessive multiplication of harmful bacteria can reduce intestinal decay, clean the intestine, promote health, and slow down aging.

The most typical symptom of maladjustment in gastrointestinal flora is constipation. How does such maladjustment cause constipation, then? Existing scientific studies tell us that food is subjected to gastric digestion, then it enters the intestines. When bifidobacteria and other beneficial bacteria in the intestines are unbalanced, food digestion and absorption decrease. Meanwhile, harmful bacteria multiply, resulting in the accumulation of corrupted metabolites in the intestines, and subsequently, the accumulation of stool or toxins. The remains of digestion wrap around the intestinal walls, hindering nutrition absorption, which leads to metabolic disorders, and thus, constipation. The cause for diarrhea is similar to that for constipation, only that they are cases of two extremities.

When our body is experiences chronic diarrhea or constipation, what should we do to restore intestinal flora balance in a short time? We can take direct supplement of bifidobacteria, lactobacilli, and other beneficial intestinal bacteria; we can also add some "bifidus factors", "prebiotics" and "probiotics", which can promote growth and regeneration of bifidobacteria.

A genetic census was carried out by German scientists, which examined fecal bacteria of European, American, and Japanese people, and revealed surprising findings. Despite racial difference, three main genera of bacteria were observed in their intestines. The first category is the *Bacteroides*, the second is *Prevotella*, and the third is *Ruminococcus*. Each individual body has these three types of bacteria among other bacteria. But in different peoples, one of the three bacteria may be dominant, or make up the majority. Thus, the bacterium that prevails in number is the bacteria of these people. One can also classify their intestines accordingly.

Skin

The surface of the skin is covered with bacteria. Every square centimeter of the skin contains roughly 10,000–100,000 bacteria. There are two most common bacteria on the skin. One is *Corynebacterium acnes*. If this clogged the pores, it would cause acne. The other is *Staphylococcus epidermidis*. This is a common skin bacterium that always gathers in groups. The number of these bacteria in the armpits can reach up to 10 million per square centimeter. Because of moisture, the belly button is a "cozy home" for bacteria.

Bacteria on the skin cannot be completely removed. Scientific statistics show that ordinary toiletries remove only surface stains on the skin. Usually, with 3 h, of showering, the number of bacteria on the skin can increase 500 times and back to its normal level.

We know that before surgery, surgeons wash their hands thoroughly with soap, and disinfect with disinfectant. Then, they put on sterile gloves for surgery. When the surgery is over, however, a lot of bacteria can be found in the sweat accumulated in the gloves. Its amount and variety is so incredible that some people jokingly call the sweat "glove soup". Where do all these bacteria come from? It turns out that, although carefully washed and disinfected, bacteria hidden in the depth of sweat and sebaceous glands of a doctor's hands cannot be easily removed. With the secretion of sebum and sweat, they constantly run to the skin surface and multiply. Bacteria multiply fast; in 3 h, 10 bacteria can turn into 5,000–40,000 bacteria, which is enough to contaminate the hands. Therefore, disinfectant that surgeons use should be able to function for long periods to constantly kill bacteria emerging from the depth of the skin and prevent the hands from being contaminated.

Bacteria are everywhere. The human body inevitably catches various bacteria, many of which are highly virulent, such as *Streptococcus*, *Staphylococcus aureus*, *Salmonella*, and *Mycobacterium tuberculosis*. Normally, harmful bacteria attach to the surface of human skin or hide inside the human body. When the bacteria multiply and exceed the defense capability of the human immune system, body resistance is low because of illness or weather changes, or a skin wound is infected, these bacteria will trigger disease. Furthermore, the normal flora we talk about is only nonpathogenic for individuals with corresponding immunity. Normal flora for one individual may be abnormal for another. For a given individual, immunity against normal flora is limited. An infection is triggered when the species' proportion or amount of normal flora change drastically, or when the flora invade normally bacteria-free places such as blood.

Bacteria in water

In the vast rivers, tranquil lakes, rough seas, gurgling brooks, and all other worlds of water lurk huge bacteria armies. As different water environments can offer different types and amount of supplies, the scale and forces of the bacteria "army" also varies.

In lakes and ponds away from human activities, there are fewer bacteria because of a limited supply of nutrition (organic substances) for the "army"; normally, each liter of water contains only a few dozen to hundreds of bacteria. Bacteria living in such environments are mainly *Pseudomonas fluorescens, Chromobacterium, Achromobacter,* sulfur bacteria, bacterial clothing, and iron bacteria.

In lakes and ponds near cities and factories, there are more bacteria because of the presence of a large number of organic substances from domestic sewage and industrial wastewater polluting the water. Each milliliter of water is able to contain up to tens of millions or even hundreds of millions of bacteria. The main types of bacteria include *Bacillus, Proteus, Escherichia coli,* and *Streptococcus faecalis,* and sometimes even *Salmonella typhi, Shigella, Vibrio cholerae,* and other pathogenic bacteria. Pathogens in water primarily come from a variety of ill human and animal's excrement. Normally, because water environment is not suitable for the growth and reproduction of pathogens, these bacteria can live only for several days to several weeks in water, and they cannot survive for a long time. However, because of the flow of water, these pathogens can still travel and cause epidemics.

In the vast expanse of the ocean, there stations a bacteria army with special skills. First of all, they are not afraid of salt. Every liter of seawater contains about 30 g of salt. Their second skill is resistance against enormous pressure. On plains, there is normally 1 atm of atmospheric pressure; but in oceans, thousands or even tens of thousands meters deep, the atmospheric pressure can shoot up to thousands or tens of thousands of atm. Bacteria in the ocean can grow and reproduce under such extreme pressure. Their third skill is that they can withstand cold conditions. About 90% of seawater is below 5 °C. The fourth skill is living on little food. There is little nutrition in deep seas. Some types of planktonic marine bacteria have adapted to low nutrient seawater; once given rich nutrition in laboratory environments, they stop growing and reproducing instead. The ocean is always undergoing dramatic changes yet constantly remaining in equilibrium, always full of vitality. Marine bacteria play an important role here. When the dynamic balance of marine ecosystem is damaged, marine bacteria, with their sensitivity and adaptability, can multiply with great speed and form abnormal microbial flora in short time, actively participating in various activities to promote the formation of a new dynamic equilibrium. However, there are also many bacteria in the sea that are harmful to the human body. For example, there is a bacterium called the sulfate-reducing bacteria (SRB). It particularly likes to live on sailing boats. As many bacteria attach to the boat, they not only destroy the boat but also cost more power for navigation.

As the saying goes, diseases enter by the mouth. As there are so many bacteria in water, we must ensure the quality of drinking water. Quality standards of drinking water include three parts. First, to prevent epidemics, drinking water should not contain pathogenic microorganisms; second, to prevent acute and chronic intoxication and potential long-term risks (carcinogenic, teratogenic, and mutagenic effects), chemical substances and radioactive materials in water should not threaten human health; and third, drinking water must ensure a good taste. As there are not many pathogens in water, and *Escherichia coli* and pathogens both come from fecal contamination of animals, one can check the number of *Escherichia coli* in water to determine the degree of water pollution by human and animal excrement, thus indirectly deduce the probability of the existence of other pathogens. According to China's drinking water health standards, the number of bacteria in each milliliter of water should not be more than 100; the number of *Escherichia coli* in each liter of water should not exceed 3.

Bacteria in soil

We often say soil is the "home base" of bacteria. Why? This is because it is very convenient for bacteria to eat and live there. One can say that a handful of soil is a colorful world of bacteria. Soil is rich in flora and fauna remains and in a variety of inorganic residues. It provides ample food for bacteria. Water and air in the soil can satisfy the needs of bacterial growth. Moreover, the pH of soil is close to neutral, and the temperature is generally constant all year round. It also has a suitable osmotic pressure for bacteria to grow. Therefore, in fertile soil, every gram contains billions of bacteria. Even in barren soil, each gram contains hundreds of millions of bacteria. There are hundreds of types of bacteria in soil. Bacilli are the largest in number, followed by cocci. There are fewer vibrio and spirochetes. Spores of some pathogens such as *Bacillus anthracis*, *Clostridium tetani*, and *Botulinum* toxin can survive for a long time in soil. Therefore, soil can easily cause wound infections.

Stationed in different soils, bacteria also develop different skills. Scientists can identify "special forces" of bacteria to serve human beings with their different skills. For example, they can find bacteria that live on oil in oil-rich areas, bacteria that decompose pollutants in discarded plastic bottles or plastic films in fields, and metal-eating bacteria in mining areas that can help people find gold.

Bacteria in the air

Nutrition and water that is required for bacteria to live is lacking in the air. Meanwhile, ultraviolet light in the sun is the "mortal enemy" of bacteria. So why are there still bacteria in the air? In fact, there are numerous tiny dusts and water drops in the invisible and intangible air, which are the "hiding place" of bacteria. Meanwhile, some bacteria or bacterial spores with high resistance to ultraviolet and dryness can survive for a long time in the air. The number of bacteria varies according to the cleanliness of the air. Generally speaking, the more the dust is, the more the bacteria. There are more bacteria in the air over land than over an ocean; more in the air over cities than in rural areas; more in the air of dirty districts than in clean places; there are most bacteria in the air of places where the concentration of humans and animals is high. Indoor air has more bacteria than outdoor air, especially in densely populated public places such as hospital wards and clinics. As dust falls naturally, the closer the air is to the ground, the more the bacteria. Droplets, dander, sputum, pus, and feces carry numerous bacteria and can seriously pollute the air. Some medical procedures such as high-speed dental drill or ultrasonic cleaning can also cause air pollution; they produce aerosols for bacteria. Bacteria on fabric surfaces fly into the air when we dress or make the bed. The cleaning and the moving of people make dust float and are the main source of bacteria in hospitals. Therefore, people living in downtown areas need to visit the suburbs and parks more often to breathe fresh air. This will be of great benefit to their health. Especially in forest parks, the air is very clean and contains fewer bacteria, an ideal place for tourism, vacations, and convalescence.

There are no fixed types of bacteria in the air. Most of the time, bacteria in soil, humans, and animals float into the air with particles and dusts. Common bacteria in outdoor air include spore-producing bacilli and pigment-producing bacteria. The most common pathogens in indoor air include *Neisseria meningitidis*, *Mycobacterium tuberculosis*, hemolytic bacteria, diphtheria bacillus, pertussis, and others. Bacteria float in the air and enter human body with respiration. Therefore, modern public health standards emphasize a safe bacteria level in the air in public places, taking it as an important indicator of air cleanliness. Scientists use PM2.5 to indicate air quality. PM2.5 refers to particles in the atmosphere with a diameter less than or equal to 2.5 μm (thinner than 1/20 of human hair). These particles are also known as inhalable particles. A higher PM2.5 measurement indicates more polluted air. Although PM2.5 constitutes only a small portion of the atmosphere, it bears great influence to air quality and visibility. In comparison to bigger particles, PM2.5 is small in size and rich in toxic and harmful material. They remain in the air for a long time and can be transported to distant places, so they have a greater impact on human health and atmospheric quality. The main sources of PM2.5 are residues of emission from daily power generation, industrial production, and automobile exhaust; most of them contain heavy metal and other toxic chemicals. Generally speaking, coarse particles of 2.5–10 μm come primarily from road dust; particles below 2.5 μm (PM2.5) come primarily from the combustion of fuels (e.g., motor vehicle exhaust and coal burning). Nowadays, we often encounter grayish surroundings. We used to think it was just cloudy, but actually it is fog caused by high levels of PM2.5 and PM10.

Do Bacteria Have "Internal Organs"?

We had a crude look of bacteria's appearance in earlier sections. But we cannot help asking: do bacteria have "internal organs"? If not, how can they grow and reproduce? If so, what do they look like? Bacterium is a microorganism. It does not have internal organs like those of animals. But it does have its own special "internal organs". If we dissect a bacterium (which is a cell), its main "organs" look as shown in Fig. 1-7. We call these organs cell wall, cell membrane, ribosomes, bacterial chromosome cytoplasm, capsule, flagella, and pili. Now, let us start from outside to inside and take a look at the "organs" of bacteria.

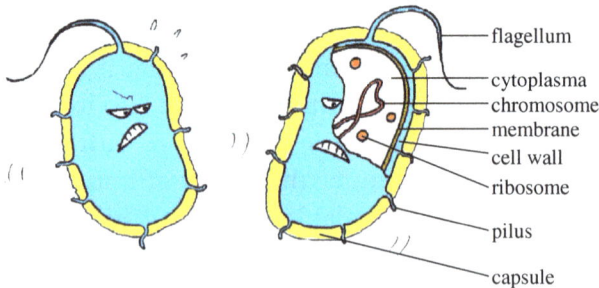

Figure 1-7:   Structure of bacterial cells.

Pili: the beard

Some bacterial surfaces have short and straight "beards" sticking out from inside the cell. These are called pili. Pili help bacteria adhere to each other and attach firmly to the mucosal surfaces of respiratory, gastrointestinal, and urogenital tracts.

Flagellum: the braid

Some bacteria have one or more curly "braids" on their surface. These are called flagella. Flagella help bacteria to move.

Capsular: the bulletproof vest

On the surface of some bacteria, there is a sticky layer of gel-like substance with varying thickness. This layer is called the capsular. It is like a bulletproof vest that bacteria wear on the outside to protect themselves from dry environments or being devoured and digested by phagocytes. Meanwhile, capsular helps bacteria invade and stick firmly to organisms.

DNA: the egg yolk

If we compare a bacterial cell to a boiled egg, then DNA, the genetic material of bacteria, is like the egg yolk in the middle of the egg. Similar to many other organisms, DNA is where genetic information is stored.

All materials that constitute life are produced under direction of the genetic information stored in DNA.

Cell wall: the eggshell

If we look at a bacterial cell as an egg, then the cell wall is like the eggshell. It is the "city wall" that protects the bacterium. Wrapped around the surface of a bacterium, the cell wall is a layer of slightly tough and flexible structure whose inside is closely attached to cell membrane. It shapes the bacterium as well as protects it.

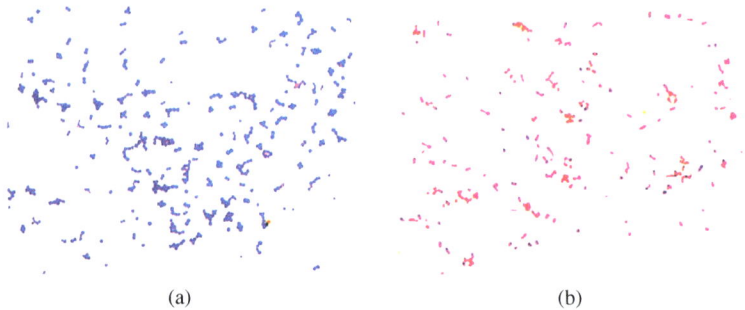

(a)   (b)

Figure 1-8:   Gram staining of bacteria. (a) Gram-positive bacteria (*Staphylococcus aureus*). (b) Gram-negative bacteria (*Escherichia coli*).

Since bacterial cells are small and transparent, to observe their shape, they must be stained. A dyed cell displays a sharp color difference with its background thus can be clearly observed under ordinary optical microscopes. In 1884, using four steps — primary stain, mordant, decoloring agent, and counterstain — Danish doctor C. Gram divided all bacteria into two categories: those turning blue-violet after staining are called Gram-positive bacteria, marked as G+; those turning red are called Gram-negative bacteria, marked as G– (see Fig. 1-8). Most pyogenic bacteria are Gram-positive bacteria, whereas intestinal bacteria are mostly Gram-negative bacteria. Different reactions to Gram stain are caused by different components and structure of the cell wall.

Whether they are Gram-positive bacteria or Gram-negative bacteria, the main component of their cell walls is a substance called peptidoglycan. These materials can weave into a mechanic web, like layers of steel frames and sheets. Then, these weblike layers are hinged by a substance called teichoic, forming a thick bacterial "city wall" — the cell wall. Although osmotic pressure is high on the inside of the cell, and water molecules from the outside, which have lower osmotic pressure, flow into the cell constantly, the cell wall can tightly wrap around the cell to prevent it from "swelling" to rupture, as shown in Fig. 1-9(a). In everyday life, when we immerse fruits in salty water to clean them, or put food in highly concentrated sugar water

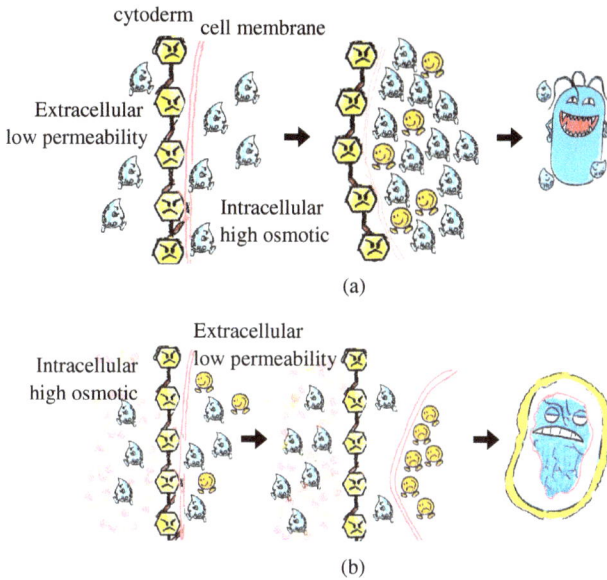

Figure 1-9: Structure and function of bacterial cell wall. (a) Water molecules can move in and out of cell walls easily. When under hypotonic environment, extracellular water molecules penetrate into the cell. The cell expands, but with cell wall wrapping around it, the cell does not burst. (b) When the cells are under hypertonic environment, intracellular water molecules move outside the cell. The cytoplasmic contracts and separates with the cell wall. Cell dies due to dehydration.

to prevent bacterial infection, the basic principle is to create a hypertonic external environment so that water molecules in the cell flow out and bacteria die of splitting from cell wall, as shown in Fig. 1-9(b).

What is different for Gram-positive bacteria is that they have simple cell walls that are thick and dense, sometimes having 20 layers. About 90% of cell wall is peptidoglycan, whereas the remaining 10% is teichoic acid. Cell wall of Gram-negative bacteria is thin and complex. It has only one to two layers of peptidoglycan, and its mechanical strength is considerably weak. Outside the peptidoglycan layer, there are also lipopolysaccharide, phospholipids, lipoprotein, and other materials that constitute the exterior cell wall.

Figure 1-10 shows the different surface structures of Gram-positive and Gram-negative bacteria.

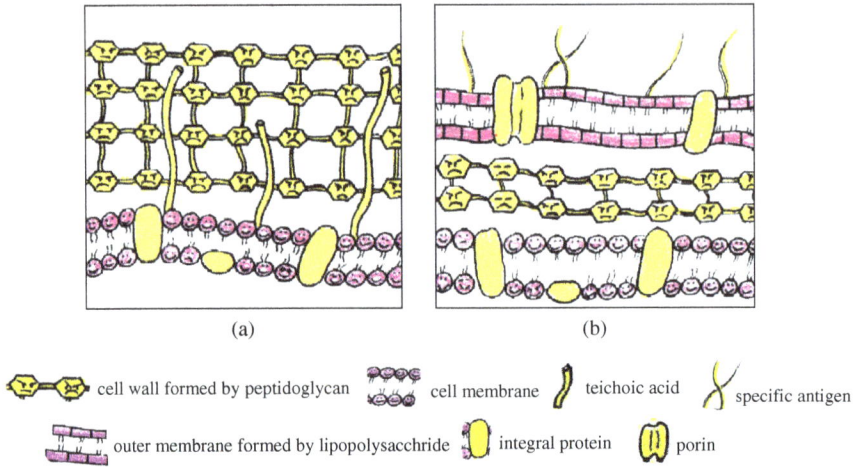

(a)   (b)

cell wall formed by peptidoglycan   cell membrane   teichoic acid   specific antigen

outer membrane formed by lipopolysacchride   integral protein   porin

Figure 1-10:   Structure of (a) Gram-positive and (b) Gram-negative bacteria.

Cell membrane: the candy wrapper

The bacterial cell membrane, also known as the cell membrane, is a layer closely attached to the interior of cell wall. It is a soft, fragile, flexible, and semipermeable film that wraps around cytoplasm, like the thin film between the eggshell and the egg. Do not underestimate this thin layer of the cell membrane.

It is the center for bacterial metabolic activities. It plays an important role in breathing, energy production, movements, biosynthesis, exchange, transfer of internal and external materials, and other activities.

Cytoplasmic inclusions

The cytoplasm is like the egg white, surrounded by a membrane. There are many transparent, gel-like particles in cytoplasm. These particles help maintain intracellular balance and restore nutrients.

Ribosome: place for protein synthesis

The ribosome consists of RNA and protein. It is where protein synthesis takes place. Ribosomes often exist in cytoplasm as free ribosome or polysome.

Spore: the dormant body

Some bacteria in their later phase of growth form a round- or oval-shaped dormant body inside the cell. This body has thick walls, little water, high resistance, and is called a spore. Spores can resist heat, radiation, chemicals, and hydrostatic pressure. Their ability to survive is the most amazing feat in the living world.

How Bacteria Reproduce?

Bacteria reproduce at an alarming rate. *Escherichia coli*, for example, when under proper conditions, can have a new generation every 12.5–20 min (certain bacterium takes longer to divide; *Mycobacterium tuberculosis*, for example, requires 18–20 h). *Escherichia coli* splits three times every hour, and each time, one is divided into eight. It can reproduce 72 generations in a day, from 1 bacterium to 4,722,366,500 trillion (weighing around 4,722 tons); after 48 h, *Escherichia coli* will have 2.2×10^{43} offspring. These bacteria weigh as much as 4,000 earths.

Of course, because of various constraints, such maddening regeneration is almost impossible. The doubling of bacteria can last only several hours instead of permanently. Regeneration consumes nutrients; toxic substances accumulate; and pH of the environment also changes. Therefore, bacteria cannot regenerate endlessly at a steady rate. After some time, regeneration slows down, and the number of dead bacteria increases. Thus, when growing bacteria in the culture, bacteria can generally reach the number of 100 million to 1 billion, even up to 10 billion per milliliter. Even so, bacteria have a much higher reproduction rate than more complex organisms.

One bacterium becoming two daughter bacteria is called binary fission breeding. This is the most common way of reproduction among bacteria. Before splitting, bacteria first stretch their body; DNA is duplicated and keeps separating; meanwhile, bacterial cell splits in the direction that is perpendicular to its long axis. Around the "equator" of bacteria, the cell membrane grows inward and forms a diaphragm, creating new cell wall at the same time. This way,

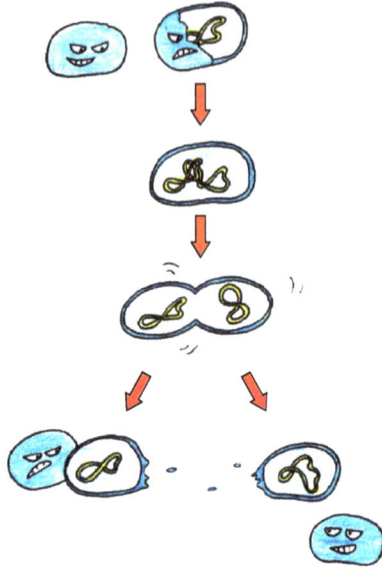

Figure 1-11:   Binary fission reproduction.

one bacterium splits into two identical daughter cells. Figure 1-11 shows binary fission.

Colony: The Bacteria Army Visible to the Eyes

In our everyday life, we may sometimes find mold in damp corners or spoiled food. This is an army — the mold colony — formed by single mold (a type of microorganism that is different from bacteria). If you observe closely, there usually are blobs that look like mucus, sometimes with different colorations — those are "colonies" reproduced by a single bacterium — the bacterial colony.

A single bacterium or small clusters of bacteria absorb nutrients from the surface of nutrient substrate. They grow and split continuously, forming a visible subcell mass with certain morphological structures surrounding the mother cell. This we call a colony. Although different bacteria form colonies with different structures, those colonies share some common features. They have a wet surface, and they are sticky, smooth, relatively transparent, and easy to

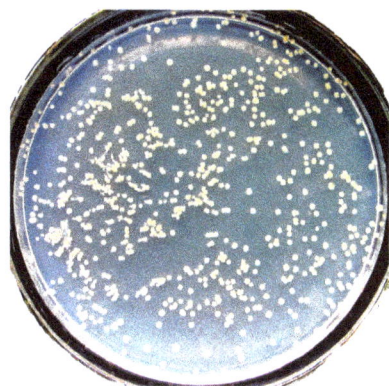

Figure 1-12: Bacteria colonies.

pick up. Their texture is even, and their back and front or periphery and center have the same color. Figure 1-12 shows bacteria colonies.

How Did Humans Discover and Understand Bacteria

From life experiences, humans learned that some "things" can rot food, and we preserve food in salted, candied, smoked, or dried form to prevent them from deteriorating, as well as increasing its flavors. However, we have encountered bacteria without getting to know them. It was not until the Dutch scientist Leeuwenhoek who discovered bacteria with self-made microscope that the sophisticated world of bacteria was revealed to us. Later, the French scientist Pasteur and the German scientist Koch started our journey to understand bacteria.

Originator: Leeuwenhoek

In 1632, Antonie van Leeuwenhoek (Fig. 1-13) was born in a small city in the east of the Netherlands called Delft. He did not receive any higher education, and he started apprenticeship in an Amsterdam cloth shop at the age of 16. Later, he opened his own small cloth shop in Delft. Back then, people examine the quality of cloth with magnifying glasses. But Leeuwenhoek was not satisfied with examin-

Figure 1-13:   Antonie van Leeuwenhoek (1632–1723).

ing his cloth with existing magnifying glasses. So he began to learn to make his own magnifying glasses.

In 1660, Leeuwenhoek got a new position as the administrator in Delft city hall. This was a leisurely job, so he had much more time to make magnifying glasses. Eventually, his glasses were achieving higher and higher magnification. Because the greater the magnification, the smaller the lens, Leeuwenhoek used two metal plates to sandwich the lens for convenience, and he attached a pointed metal stick to the front of the lens, so that things could be observed at the point. There was a screw knob to adjust the focus. He made up to 500 microscopes in his lifetime. Although they are more magnifying glasses with high powers rather than microscopes in its strict modern sense, but Leeuwenhoek observed raindrops, sewage, blood, pepper water, corrupt substances, alcohol, butter, hair, semen, muscle, tartar, and many other substances. He also became the first person to have observed the world of microorganisms, which he called "tiny creatures". He wrote more than 300 letters to scientists in the Netherlands and other countries. Some letters were translated by his friend Regnier de Graaf. de Graaf believed that more people should know about Leeuwenhoek's work, so he urged Leeuwenhoek to contact the Royal Society of London for Improving Natural Knowledge. In 1676, Leeuwenhoek wrote a lengthy letter to the secretary of the Royal Society, reporting his observations of the past 20 years. He also included a picture of tiny creatures that people had never seen before — spherical, rod-shaped, and spiral-shaped bacteria and protozoa. Leeuwenhoek's discovery was a sensation, and his letters in Dutch were translated into English or Latin and published in the journal of the Royal Society of London. Leeuwenhoek wrote 190 letters in total to the Royal Society and donated 26 microscopes. In 1680, he was elected a member of the Royal Society of London; in 1699, he was appointed the correspondent of the Academy of the Sciences in Paris; in 1716, he received a silver medal from the University of Leuven. Apart from observations through his microscopes and reporting his observations, Leeuwenhoek had no other hobbies. At the age of 90, 36 h before the end of his life, he was still writing to the Royal Society of London.

Leeuwenhoek is the founding father of bacteriology. However, in the next 200 years after Leeuwenhoek's discovery, human knowledge of bacteria was still limited to describing their forms and shapes. It was not until people used microscopes with higher magnification to reobserve Leeuwenhoek's "tiny creatures" and realized that bacteria can both cause disease and be beneficial did the greatness of Leeuwenhoek's contribution be fully appreciated.

Founding father of microbiology: Pasteur

Louis Pasteur (Fig. 1-14) was born in Dours, France, in 1822. His father was a cavalry officer under Napoleon. Since childhood, Pasteur had been determined to be a knowledgeable person. He set strict requirements for himself during school, trying to achieve perfection in every subject. In the summer of 1843, Pasteur entered Ecole Normale de Paris. Under the influence of then chemistry master J.B. Dumas, Pasteur devoted himself to the world of chemis-

Figure 1-14:   Louis Pasteur (1822–1895) in his lab.

try. Pasteur was conscientious, devoted, and did not work for fame or money. His professor believed such attitude to be the best reward for a teacher. At 25, Pasteur earned his PhD and remained a teaching assistant at his school. At 26, he discovered the principle of rotation, which at that time remained an unsolved question for many scientists. This principle opened new possibilities to research in stereochemistry, to which he was the founder.

In 1853, Pasteur became professor in Universite de Lille and devoted himself to teaching and tackling problems faced by local industries. In 1873, he was elected member of the Academie Nationale de Medicine. In 1888, he established in Paris the Institut Pasteur, an institute specialized in curing diseases. At his death in 1895, Pasteur was honored as a national hero and received a state funeral. Pasteur devoted his entire life to scientific research. He paved way for new fields of microbiology such as fermentation, bacterial culturing, and vaccines. Figure 1-14 shows a painting of Pasteur (1822–1895) conducting an experiment in his lab.

Pasteur was one of the most accomplished scientists of the 19th century. He was acknowledged as the "father of microbiology" and the "most flawless man entering the kingdom of science". "Will, work, and success are the three elements of life. Will opens the door to your career; work is the entering path; at the end of the path stands success to celebrate your hard work — so long as there is a strong will and hard work, there must be a day of success", this was Pasteur's famous motto for success.

Life comes from life

Since ancient times, people have been exploring: where does life come from? There are records in both Eastern and Western civilizations: from rotten meat come maggots, and from rotten grass come fireflies; Greek goddess of love Aphrodite was made of foams from the sea; if one put bread crumbs in a mouse cage, mice will jump out. The idea is that life occurs "naturally" from things that are not living.

The first critic of the "spontaneous generation" theory was an Italian doctor, Redi. He conducted an experiment: he cut a piece of fresh meet into two, and placed them separately in two clean containers. One container was covered by cheesecloth, and the other was left uncovered. After a while, Redi discovered that meat in the uncovered container had maggots on it, whereas meat in the covered container did not have them. Therefore, Redi concluded that, instead of coming directly from rotten meat, maggot eggs must have been laid by flies coming from outside the container. But supporters of spontaneous generation raised objections: although animals and plants could not generate spontaneously, microorganisms could. Figure 1-15 shows Redi's much questioned experiment.

Flies come into the bottle, lay eggs on the meat, hatch grub then meat deteriorates	Flies can not come into the bottle covered with gauze to lay eggs, meat does not deteriorate

Figure 1-15:   Redi's experiment questioning "grub from rotten meat".

In the meantime, Pasteur conducted a "curved-neck flask" experiment (Fig. 1-16), which further refuted spontaneous generation theory. He put nutrient solution (such as broth) into a flask with curved neck. The top of the flask was open and air could come inside unhindered. Bacteria in the air were then trapped and accumulated at the curve of the neck instead of entering the flask. Pasteur boiled the broth on fire to kill spores of bacteria and other microorganisms. He then left the broth stationary. In a long time afterward, there would be no bacteria in the flask. This was because when air flew through the neck, spores of bacteria and other microorganisms were cleansed. They settled at the bottom of the curved neck and could not make contact with the broth. If the curved neck is inclined and air entered without any filtering, the nutrient solution would soon be full of bacteria.

Pasteur's experiment proved that like other organisms, microorganisms could not generate spontaneously from nutrient solution either. Instead, they came from the microorganisms (spores) already existing in the air. The "spontaneous generation" theory that had influenced the world for thousands of years came to an end. Since then, the door to microbiology studies was opened.

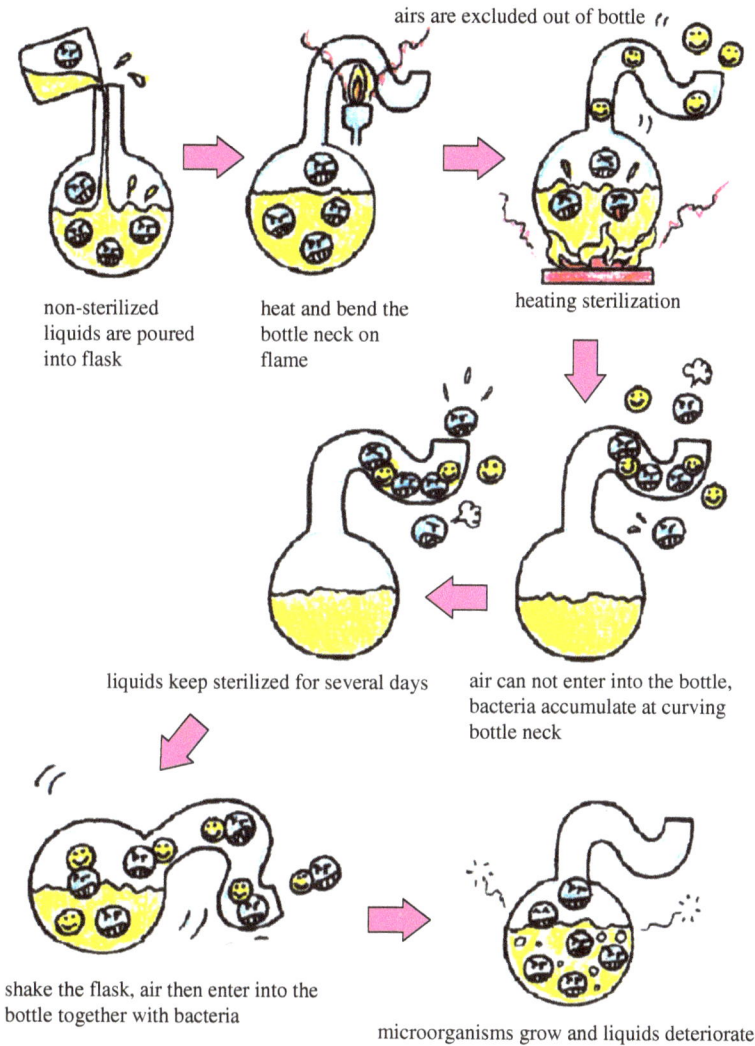

airs are excluded out of bottle

non-sterilized liquids are poured into flask

heat and bend the bottle neck on flame

heating sterilization

liquids keep sterilized for several days

air can not enter into the bottle, bacteria accumulate at curving bottle neck

shake the flask, air then enter into the bottle together with bacteria

microorganisms grow and liquids deteriorate

Figure 1-16: Pasteur's "retort" experiment.

Secret of sour wine

France leads the industry of beer and wine production in Europe. But wineries often encounter one problem: wines turn sour. An entire barrel of aromatic wine turns into a sticky and sour substance that makes people grimace. These barrels have to be thrown away, a painful lost to wine producers. Some even went bankrupt because of this.

In 1865, a city brewery in Lille, France, asked Pasteur to help solve the problem of sour wine. As Pasteur was then already a famous chemist, the brewery owner hoped he could add some chemicals to prevent wine from turning sour.

Pasteur did not solve the sour wine problem with chemicals; instead, he examined under microscope the difference between normal wine and souring wine.

He discovered that in normal wine, there was only one kind of round and big yeast, whereas in souring wine, there is another type of thin and long bacteria (Fig. 1-17 shows how yeast and this bacteria looked under an electron microscope). With these bacteria in normal wine, the wine turned sour. Therefore, Pasteur put sealed wine bottles in water heated to different temperatures. He wished to kill the bacteria without harming the taste of wine. After numerous experiments, he finally found the easiest and most effective method: put the wine in 50–60°C environment for half an hour; bacteria that turned the wine sour would be killed, and the wine remained as aromatic as before.

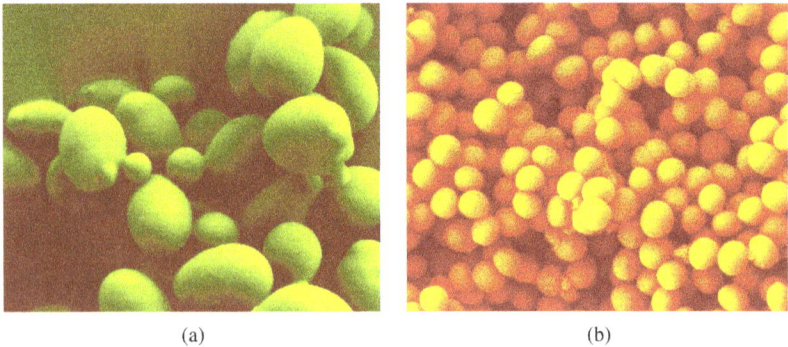

(a)   (b)

Figure 1-17: Yeast and bacteria under electron microscope. (a) The small ball on the big ball of yeast. This is the reproduction process of yeasts. Scientists call this "budding". (b) Dividing bacteria. It is smaller in sizes than yeast.

The method of killing bacteria with mild heat was since then called pasteurization. Until today, the milk we drink every day is still preserved with pasteurization.

Why do silkworms get sick?

After solving the problem of sour wine, Pasteur gained great reputation in France. At that time, the silk industry in southern France was facing the crisis of sharp decrease in production. The reason was that a disease called pebrine caused massive death of silkworms. Again, people turned to Pasteur. Pasteur had no previous experience with silkworms. But at the thought of the hundreds of millions of francs lost due to pebrine, Pasteur did not hesitate. In 1865, appointed by the Minister of Agriculture, Pasteur took his microscope to southern France where the industry was seriously damaged. After years of work, Pasteur discovered that the source of pebrine was a tiny and oval brown particle. Healthy silkworms got sick after eating mulberry leaves brushed with this particle. Sick silkworms could also infect healthy ones through excrement. Pasteur told the farmers that they must burn all the infected silkworms and their eggs together with mulberry leaves. New silkworms should only be incubated from healthy worms. Following Pasteur's instruction, the farmers disposed of infected worms and detained the further spread of disease. Pebrine was prevented. Pasteur saved the silk industry of France and was awarded by Napoleon the Third.

Pasteur's experiment proved that the infectious disease among silkworms was due to bacterial invasion. This success turned Pasteur's research interest to infectious diseases. According to him, if silkworm diseases were caused by microorganisms, they should also be the main cause for human diseases. This is the pathology theory of diseases. Pasteur's vision was praised as the greatest discovery in the history of medicine.

Turning "evil" bacteria into vaccines

Pasteur successfully researched and discovered vaccines for cholera, anthrax, and rabies, opening a new era of humans conquering infec-

Figure 1-18: Robert Koch (1843–1910).

tious diseases. He laid the foundation of immunology, an important branch of science today. His place in history is unprecedented. For more, read Less "Poisonous" Enemies are our Friends in Chap. 4.

Pioneer of bacteriology: Koch

Robert Koch (Fig. 1-18) was born in December of 1843, in Claudia Stelle near Hartz, Germany. In 1866, he graduated from the Medical Department of University of Göttingen. He then went to Berlin for 6 months of chemical research. In 1867, he became a resident doctor in Hamburg. During the Franco-Prussian War, he was the army surgeon. After the war, he opened a clinic in a town in eastern Prussia.

During his practice, he built a simple laboratory at home. Without equipment, access to libraries, or contact with other researchers, Koch started his research and accomplished many "firsts" in microbiology:

In 1876, he proved in a public presentation that anthrax bacillus was the pathogen of anthracis, which proved that certain bacteria caused certain diseases.

Around 1881, he created microscopic photography technology and staining techniques to observe bacteria under microscopes.

In 1882, he invented the technique of separating and purifying microorganisms with solid agar medium.

In 1882, he isolated *Mycobacterium tuberculosis.*

In 1883, he isolated *Vibrio cholerae.*

In 1884, he established guidelines for identifying pathogenic micro-organisms.

During 1891–1899, he discovered that rat flea is the media of plague, and trypanosomiasis is spread by tsetse fly.

In 1905, because of his contribution to tuberculosis research, Koch was awarded the Nobel Prize in Physiology or Medicine.

All these accomplishments demonstrated the breakthroughs Koch had introduced to science and made him the tycoon in bacteriology. On May 27, 1910 at the age of 67, Koch died of heart attack in Baden, Germany.

Pathogens "hunter"

Since ancient times, plague, typhoid, cholera, tuberculosis, and many other horrible diseases have taken many human lives. To overcome these diseases, humans needed to first understand their causes. Pasteur discovered sheep anthracis bacteria because of the presence of anthracis bacteria in the blood. However, he did not connect the microorganism with the disease. As a doctor, Koch had comprehensive knowledge of human body, which Pasteur lacked. Meanwhile, Koch had superb experiment skills and the quality of perseverance and patience as a scientific researcher. It was Koch who first discovered that infectious diseases were caused by pathogenic infections. He was the founder and pioneer of pathogenic bacteriology.

On the basis of Pasteur's work, Koch started a systematic study of anthracis bacteria. For three full years, Koch was fully devoted to finding out the cause of anthrax. He found pathogens of anthrax in cattle spleen and planted the pathogens in a mouse so the mouse was infected with anthrax. He then rediscovered the same bacteria as he had found in the cattle. In 1875, Koch successfully confirmed that *Bacillus anthracis* was the cause of anthrax. This was the first time that humans scientifically proved that a certain microorganism was the cause for a certain disease (Pasteur found the vaccine for anthrax in 1881).

In addition, Koch successfully cultured anthracis bacteria outside of an animal body with the serum at the temperature of the cattle's body, and he discovered the life cycle of the bacteria. Anthracis bacteria can survive for a long time outside of an animal's body, and their spores can grow into bacteria and infect other animals.

In 1882, Koch discovered the pathogen that causes tuberculosis. He successfully separated *Mycobacterium tuberculosis* in solid serum medium. He replanted the bacteria in guinea pigs; they then developed tuberculosis. In 1883, Koch isolated and cultured *Vibrio cholerae* in India.

"Koch's postulates"

On the basis of his experience of isolating pathogens, Koch came up with the famous "Koch postulates": a pathogenic microorganism must be found in abundance in all organisms with the disease, but it should not be found in healthy organisms; the pathogenic microorganism must be isolated from a diseased organism and grown in pure culture; the cultured microorganism should cause disease when introduced into a healthy organism; the microorganism must be re-isolated from the inoculated and diseased experimental host and identified as being identical to the original specific causative

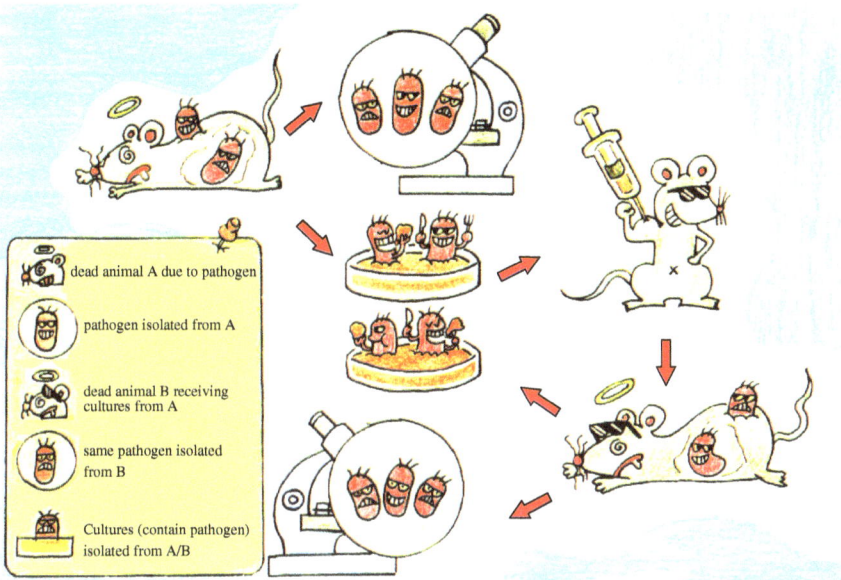

dead animal A due to pathogen

pathogen isolated from A

dead animal B receiving cultures from A

same pathogen isolated from B

Cultures (contain pathogen) isolated from A/B

Figure 1-19: Koch postulate.

agent. The process of establishment of Koch postulates is shown in Fig. 1-19.

These postulates are applied to identify whether a certain bacterium is a pathogen. The method is still used today and has made great contributions to human kind. With the guidance of these postulates, the period between 1870s and 1920s was called the golden age of pathogenic discoveries. In 1883, diphtheria was found; in 1884, typhoid bacillus was found; in 1894, the plague bacillus was found; and in 1897, dysentery was found. Until 1990, 21 pathogens were discovered in the short 21 years. "With the right methods, discoveries will come like ripe apples falling from the tree". And it was Koch who discovered this right method.

Chapter Two

Friend or Foe?

If I asked you, do you like bacteria? Are bacteria your friend or foe? Probably the first word that comes to your mind is "disease". Indeed, many terrible diseases such as plague, cholera, and anthrax have their inseparable cause in bacteria. However, you may not be aware that not all bacteria have fierce faces, sharp claws, and a malicious nature. Many bacteria are "gentle and lovely". In addition, under certain circumstances, those malicious bacteria can become our "intimate ally" after being reformed by scientists. They can be the crucial weapons for us to defeat disease and protect health. Still many bacteria make great contributions to industry, agriculture, and environmental protection. There are two sides to bacteria. When they are positioned in the proper place, they can be our friend; once they walk in the wrong place, they become our enemy. Therefore, bacteria are both our friend and our foe (Fig. 2-1).

Bacteria are Our Beloved Friends

Friendly coexistence with the human body

When still inside the mother's womb, a baby is "clean" and unaffected by bacteria. This is because the placenta is a "natural barrier" that blocks bacteria from entering the baby's body. However, once the baby comes out of the womb, his/her bond with bacteria is established. Bacteria start entering our body in various ways, and they accumulate, station themselves, and multiply. Generations of bacteria grow and age inside the human body, never leaving for

Figure 2-1:   Bacteria are friends and are foes as well.

their entire lives. Will these "aliens" take over the host, or even "offend" him/her, thus causing harm to the human body?

Generally, except in internal organs, blood vessels, and the lymphatic system, bacteria are present in all other human organs such as skin, respiratory tract, gastrointestinal tract, genitourinary tract, and other cavity organs connected to the outside. These bacteria are not only harmless but also beneficial to humans. They form a mutually beneficial relationship with the human body and fight alongside the immune system to resist "foreign invasion". Scientists call the bacteria that get along well with the human body "normal flora". Normal flora maintains a balance with the human body; there is also check and balance among floras. Normal flora has great influence on nutrition, immunity, growth, metabolism, tumorigenesis, and aging. We have discussed the related topics in Chap. 1.

However, under certain circumstances such as long-term use of antibiotics or long-term suffering from chronic wasting diseases, such as malignant tumors, tuberculosis, chronic atrophic gastritis, severe trauma, burns, systemic lupus erythematosus, chronic suppurative infection, chronic blood loss, and other similar diseases that consume energy excessively and break the energy balance of the organism, the original balance between the flora and the human body is damaged. Harmless flora can turn into harmful pathogens and can lead to possible damage or disease to human bodies. Normal flora also moves around. Once moved to a wrong place, it will become a pathogen. For example, during dental extractions or tonsillectomy, *Streptococcus viridans*, bacteria that are common to respiratory tracts, emerge into the blood and settle on abnormal heart valves, which can cause subacute endocarditis.

Credits in modern industry

Antibiotics

There are many interesting phenomena in nature. Some microorganisms, for example, help each other and live interdependently in an entity. This is called "mutualism". However, some microorganisms fight each other, one producing certain substances to restrain or kill the other. This is called "antagonism", which is especially common among microorganisms. Microorganisms with antagonistic capabilities are called antagonistic bacteria. Antibiotics are the weapons that antagonistic bacteria use to battle with other microorganisms.

The first antibiotic used in clinical treatment of bacteria-infected disease was penicillin. Although it was produced from molds, the blooming scientific research based on its discovery reveals that some bacteria are also "masters" at producing antibiotics. Bacitracin produced by *Bacillus subtilis* and polymyxin produced by *Bacillus polymyxa*, for example, can interact with cell walls and led useful substances to flow out of the cell, thus killing the cell. Meanwhile, actinomycetes in microorganisms produce the most types of antibiotics. Most commonly in clinical use, streptomycin, gentamicin, and kanamycin are called aminoglycoside antibiotics; erythromycin,

Figure 2-2:   Powerful weapon, antibiotics.

spiramycin, and midecamycin are called macrolide antibiotics; tetracycline, oxytetracycline, and chlortetracycline are called tetracycline antibiotics; and daunorubicin and doxorubicin are called anthracycline antitumor antibiotics. These are all weapons created by actinomycetes to defeat the enemies. How are these weapons used against the enemies is illustrated in the chapters 4 and 5. Figure 2-2 shows the powerful weapon, antibiotics. Figure 2-3 shows the process from isolating microorganisms that can create antibiotics to the industrial production of antibiotics.

Vitamins

Vitamin is a necessary substance to maintain human activities and is also an active substance that preserves health. Although little in amount in the human body, vitamins are crucial in growth, metabolism, and development of the body. Lack of vitamins can hinder the metabolism of the organism. Lack of different vitamins can cause different diseases: Lack of vitamin C (aka: ascorbic acid) leads to diseases such as scorbutus; lack of vitamin B_2 (aka: riboflavin) leads

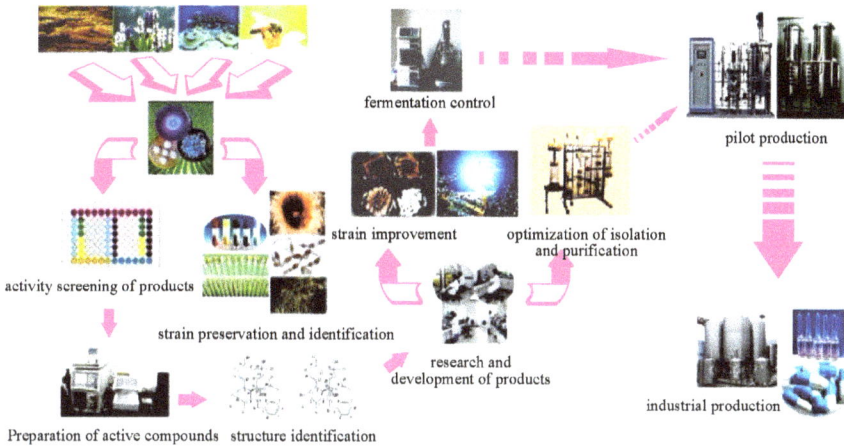

activity screening of products

strain preservation and identification

Preparation of active compounds   structure identification

fermentation control

strain improvement

optimization of isolation and purification

research and development of products

pilot production

industrial production

Figure 2-3: From isolating microorganisms that can produce antibiotics under various circumstances to the industrial production of antibiotics. (From *Introduction to Bioindustry* by the same author.)

to diseases such as angular cheilitis; lack of vitamin D leads to rickets. Therefore, vitamins are crucial medicines for clinical use. They are also used in husbandry and the feed industry.

The process that produces vitamin C from bacteria is called two-step fermentation. The first step uses *Gluconobacter melanogenus* (or *Acetobacter suboxydans*). The bacteria are cultured and expanded through second-level seeds. When quality of the seed liquid reaches the standard for seed transfer, the bacteria move to fermentation media containing sorbitol, cornstarch, phosphate, calcium, and carbonate. Fermentation occurs at 28–34 °C. Sorbitol can be constantly added during fermentation, and yield rate can reach 95%. When sorbitol makes up 25% of the media, fermentation can still continue. When fermentation is over, the broth will be pasteurized in low temperature to attain sterile broth with sorbitol. This will be the raw material for the second fermentation. In the second step, *Gluconobacter oxydans* (or *Pseudomonas*) will be cultured and expanded through second-level seeds. When seed liquid reaches the standard, the bacteria will be transferred to fermentation media containing broth from the first step. It will be incubated for 60–72 h under 28–34 °C. Finally, the broth concentrates, and vitamin C is obtained after chemical conversion and refinement.

Amino acids

Amino acid is an important component of organisms. It can be widely used in medicine, food, feed, chemical industry, cosmetics, and health products. Today, there are more than 100 types of amino acid for medical use and amino acid derivatives. In clinical situations, compound preparations with amino acids arranged in certain combination and proportion are often used to improve nutritional status of the patient. This will increase opportunities of curing and promoting health. Elemental diet is no-residue diet that contains amino acids and sufficient nutrients. It can be easily absorbed with little or no digestion. It also can be provided to patients who are weak or lack protein because of various reasons, as well as to patients who cannot digest protein. Lysine, threonine, and tryptophan are often added to wheat flour to fortify the food. Monosodium glutamate and aspartate increase the taste of food.

In 1956, *Corynebacterium glutamicum* was isolated in Japan and used in industries to produce glutamate. This triggered a great development in amino acid. By far, there are more than 10 types of amino acid that can be produced through bacterial fermentation. Normal bacteria cannot accumulate and produce amino acid in excess. To obtain amino acid in a large quantity, humans must transform microorganisms. Their normal metabolic activities need to be destroyed to create productive bacteria.

Bacteria that produce amino acids include *Corynebacterium, Brevibacterium, Brevibacterium Lactofermentum, Bacillus brevis, Serratia marcescens* and *Corynebacterium crenatum*. Apart from glutamic acid, they can also produce lysine, phenylalanine, proline, threonine, histidine, tryptophan, valine, methionine, leucine, and isoleucine.

Organic acids and organic solvents

The sour taste of oranges, grapes, and lemons is the result of the various acid substances they contain. These acid substances are called organic acids. Some bacteria can produce organic acids. *Acetobacter* and *Gluconobacter*, for example, can produce vinegar; *Lactobacillus* and *Leuconostoc* can produce lactate; *Arthrobacter paraffineus* and *Corynebacterium* can produce citric acid; and Corynebacteriaceae and *Pseudomonas* can produce gluconate.

Organic solvents such as ethanol, butanol, and glycerol are important raw materials for many industrial products. Some are also important fuels. *Clostridium* can produce butanol through fermentation.

Nucleotides

Mostly, mass industrial production of nucleotide includes guanylate and inosine. They are more than 10 times more savory than monosodium glutamate. Bacteria that produce inosine include *Bacillus subtilis, Bacillus pumilus,* and *Brevibacterium ammoniagenes*; bacteria that produce guanylate include *Corynebacterium glutamicum* and *Brevibacterium ammoniagenes. Arthrobacter ramosus* and wax-dissolving micrococcus can produce cyclic adenosine acid; *Sarcina lutea* produces flavin adenine dinucleotide (FAD); *Corynebacterium glutamicum* and *Brevibacterium ammoniagenes* produce nicotinamide adenine dinucleotide (NAD); *Brevibacterium ammoniagenes, Bacillus,* and micrococcus produce coenzyme A.

All bacteria have the ability to synthesize nucleotides. Generally, when the amount of nucleotides reaches a certain level, bacteria have a mechanism to inhibit nucleotide synthesis, thus maintaining balance between nucleotide synthesis and decomposition. To produce nucleotides, it is necessary to break this balance so that nucleotides can continuously accumulate in the culture medium.

Enzymes

Cows eat grass but produce milk. Such an amazing process is attributed to bacteria living in cows. These bacteria can produce "enzymes" that process grass and other food inside the cow's stomach into milk. Among them, the most able "craftsmen" are the cellulases. Cows feed primarily on rough grass, which contains cellulose in abundance as it is often the component of cell walls. Cellulase enters the cow's digestive system with the feeds. Cellulose is decomposed during the rumination process and produces nutrients that are more easily absorbed by cows — monosaccharides and small molecules of polysaccharide enzymes. Then after a series of complex metabolic processes, these become high-protein milk. Isolating the different enzymes produced by bacteria allows various zymins to be produced. Many enzymes from animals, plants, or microorganisms can be used in production of zymins. Compared to animals and plants, acquiring enzymes through bacterial fermentation yields high production and stable quality, and it allows for scaling up production. Figure 2-4 shows how cellulase produced in the cow converts grass into milk.

Although bacteria can produce various enzymes, only a small portion of them can be converted into zymins at the moment. For example, *Bacillus subtilis* and *Bacillus licheniformis* can produce amylase, protease, and lipase; and *Escherichia coli* can produce asparaginase. Amylase is used in industries for hydrolysis of starch into glucose and other sugars; in medicine, it is used to produce "enzyme tablets" together with protease and lipase. These tablets can treat indigestion. Proteases and lipases can also be added to detergents to clean fabrics contaminated by protein or grease. Asparagine produces aspartic acid

Figure 2-4:   Rumen bacteria convert grass into milk.

and ammonia. Patients with acute lymphoblastic leukemia lack asparagine synthetase, so they cannot synthesize asparagine. If asparagine is injected into the blood, it can affect asparagine supply in tumor tissues. The growth of tumors can be controlled since protein synthesis is deterred.

Microecological preparations

The so-called microecological preparations are preparations of viable and/or dead bacteria that maintain flora balance of the human body or improve immunity. It is also known as probiotics. The most used bacteria in microecological preparations include *Bifidobacterium, Bacillus, Lactobacillus acidophilus, Enterococcus, Lactobacillus bulgaricus,* and *Streptococcus thermophilus.* The New Regulation of Resources and Food in China explicitly states 31 types of bacteria that can be used in food. These include 16 bacteria from the Bifidobacteriaceae family, 14 from the Lactobacillaceae family, and 1 from the Streptococcaceae family. Meanwhile, the new regulation also appointed 34 zymins that can be used as food additives, including papain, glucoamylase, pectinase, lipase, and protease.

Bifidobacterium is a Gram-positive bacterium. It exists mainly in human intestines (colon). Each gram of feces contains up to 10^{10} bifidobacteria. *Escherichia coli* is the predominant bacteria in the intestines of newborn babies. After 6–8 days of birth, a bifidobacteria-dominated colony is established in the intestines. The main mechanism of microecological preparations is as follows: through normal metabolism of beneficial bacteria, lactic acid, acetic acid, and propionic acid are produced to lower the pH of internal environment of the organism; hydrogen peroxide is produced to kill potential pathogens; some metabolites are produced to decrease the concentration of ammonia and amines in the intestines; enzymes are produced to facilitate decomposition; vitamin B is synthesized and antibacterial substance is reduced; and nonspecific immunity regulatory factors are produced to increase activities of antibody and macrophage.

Microecological preparations regulate flora balance in the intestines. They promote growth of normal bacteria such as bifido-bacteria, lactobacilli, and bacteroides. They deter growth of harmful bacteria such as *Escherichia coli*. They also modify the imbalanced microecology of the digestive system to restore micro-organisms in the intestines gradually to normal. They fall into the category of ecological prevention and do not produce side effects as antibiotics do.

Apart from microecological preparations, another substance called prebiotic can also be used clinically as health-care products to regulate gastrointestinal functions. Prebiotic is officially called functional oligosaccharides, namely functional sugar. It refers to carbohydrates that cannot be completely digested by human body, including stachyose, raffinose, oligofructose, soybean oligosa-ccharide, xylooligosaccharide, and galactooligosaccharide. Oligosaccharides are composed of 2–10 monosaccharides and exist in more than 30,000 types of natural plants, including onions and garlics. As there is no enzyme for hydrolysis of these oligosa-ccharides in the human gastrointestinal system, they cannot be digested and absorbed. Instead, they enter directly into the large intestine and become food to beneficial intestinal bacteria such as *Bifidobacterium* and *Lactobacillus*. They are the multiplication factor for these beneficial bacteria. Thus, the function of prebiotics is to selectively stimulate the growth of beneficial bacteria in the intes-tines to regulate gastrointestinal functions.

Adding oligosaccharides in foods can increase the number of beneficial intestinal bacteria and decrease the number of harmful ones. Today, famous dairy producers around the world are succeed-ing in adding prebiotics to milk powder, which solves the problem of infant prebiotics supply. Some infant formula contains both prebiotics and probiotics. This can be found in the nutrition table. Figure 2-5 lists some microecological preparations used as medicine or health-care products.

By far, not only there is a dazzling array of microecological preparations, but also there are many preparations used for poultry, aquaculture, as well as husbandry of pigs, sheep, and cows. These

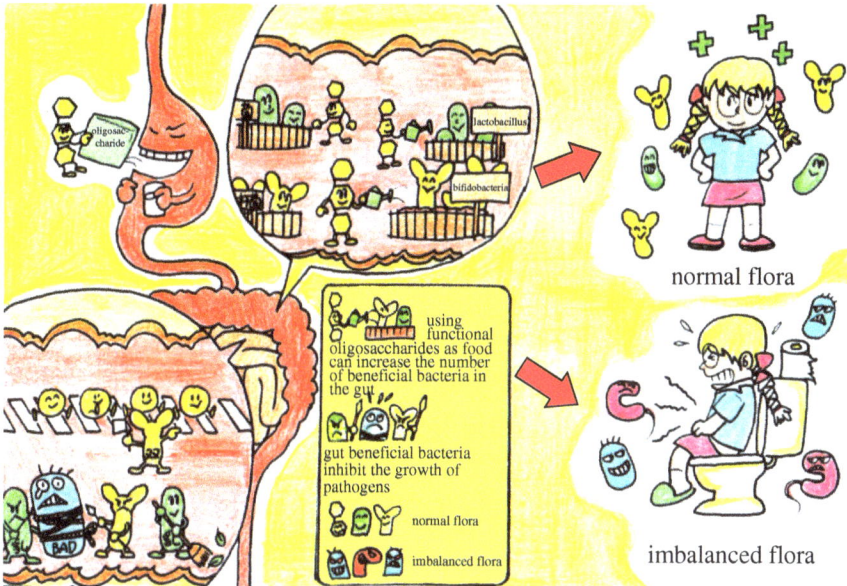

Figure 2-5: Probiotics as medicine or dietary supplements.

preparations can improve the animal immunity and increase utilization rate of feeds.

Biological pigment

Biological pigment is a chemical substance of the biological cell. Because of its special chemical structure, biological pigment can reflect, interfere, scatter, or absorb light. There are various types of biological pigments, and they are used in many ways. In industrial productions, many pigments are obtained from plants and microorganisms, including anthocyanins, carotene, indigo, lycopene, chlorophyll, Monascus pigment, and melanin. There are different types of pigment cells in the skin, fur, and eyes of animals. By far, natural pigments are mainly obtained from plants or through microorganism fermentation. Compared to artificial pigments, they are safer and brighter in color. They also possess high nutrition and pharmacological values.

Petroleum exploration

Petroleum is formed from the remains of ancient animals, plants, and microorganisms deeply buried in the earth, after a long transformation under high pressure and temperature. Petroleum is the "blood" of industry. But as it is buried deep in the earth, how can we find it? Sometimes, considerable labor, materials, and money are spent without exactly identifying the distribution range of petroleum. Here, bacteria can be said to have an inexplicable bond with petroleum. Petroleum is composed of various organic compounds, of which the majority is a carbon and hydrogen compound called hydrocarbon. Although petroleum is buried deep, there are always some hydrocarbons coming up to the earth's surface through the gaps in rock formations. Gas components in petroleum can also leak to the surface. Some bacteria feed on petroleum. Therefore, when explorers detect a great amount of such bacteria in a place, they know there is probably petroleum. On the basis of the quantity of bacteria detected in the sample, they can also predict the quantity of petroleum and gas in reserve. Bacteria in the kingdom of microorganisms that feed on hydrocarbons in petroleum include methane- and ethane-oxidizing bacteria. They can be the "guide" for petroleum explorers.

Petroleum extraction

Why use bacteria to extract petroleum? How do bacteria extract petroleum? What are the advantages of using bacteria? These may be the mysteries occupying your mind.

When a new petroleum field is developed, petroleum is thick and abundant, and underground pressure is high. At that point, petroleum may easily "flow out". After some years, underground pressure becomes lower, and water is inserted underground to propel the petroleum, which is then drawn up by a machine. Still later, inserting water underground will not help. Petroleum becomes scarce, and newly injected water often runs in the same direction of its old path. Thus, it becomes difficult to extract petroleum. Consequently, scientists adopt a chemical method for extraction, namely injecting polymers into the petroleum layer. These polymers can block the old path of water flow so that water runs in other directions where petroleum is pushed out. Although such chemical method proves to be effective, the blocked "old paths" can almost never be opened again, and petroleum trapped inside can no longer be extracted. Thus, generally speaking, the longer the extraction has occurred, the more difficult it is to extract petroleum in the field. However, one can inject bacteria and its nutrition source into the underground petroleum layers and let bacteria grow and multiply. This way, in the complex world of 1000 m underground, metabolism of bacteria can directly interact with crude petroleum. Macromolecular chains are cut into small molecular chains. The physical properties of petroleum are changed, and its viscosity is decreased. Thus, petroleum can flow more easily between the gaps in the earth formations. On the other hand, these bacteria can decompose wax and other substances in the "old paths", so that the formerly blocked places are cleared. Some bacteria can even block the "old paths" or large pores produced by habitual flows of water with metabolites. Afterward, more petroleum can be extracted by injecting water to raise pressure. Figure 2-6 is an illustration of using bacteria in petroleum exploration and extraction.

Metal smelting

Ores of many important metals such as copper and iron are buried deeply underground. To move them to the surface requires considerable labor. On the one hand, digging a tunnel from hundreds meters deep to the surface and carrying ores up is dangerous and tiring; on the other hand, digging these tunnels also requires lots of materials and money.

Today, scientists have cultured many bacteria. They let these tiny creatures surround the seam with water rapidly. Part of the bacteria such as *Thiobacillus ferrooxidans* and polysulfide bacilli oxidize sulfur or sulfide into sulfuric acid; the other bacteria such as *Thiobacillus ferrooxidans* will oxidize ferrous sulfate into ferric sulfate in the presence of sulfuric acid. Bacteria gain energy from these chemical

Figure 2-6:   Bacteria and petroleum exploration and extraction.

reactions and continue to grow and multiply. Ferric sulfate produced through oxidation further oxidizes chalcocite and pyrite into ferrous sulfate and cupric sulfate. In the meantime, ferric sulfate itself is reduced to ferrous sulfate. This way, copper ions in the ore leach out, and the liquid flowing from the seam will contain large amounts of cupric sulfate. It only needs adding some iron in the liquid to obtain copper (see Fig. 2-7). Apart from copper, bacteria are also industrially used to extract uranium and gold.

Contributions to savory food

Vinegar

Vinegar is a condiment containing acetic acid. It can increase appetite and improve digestion and is essential to our diet. Apart from 3% to 5% acetic acid, vinegar also contains various nutrients and savory factors such as amino acids, organic acids, carbohydrates, vitamins, alcohol, and esters. Therefore, it has unique color, aroma,

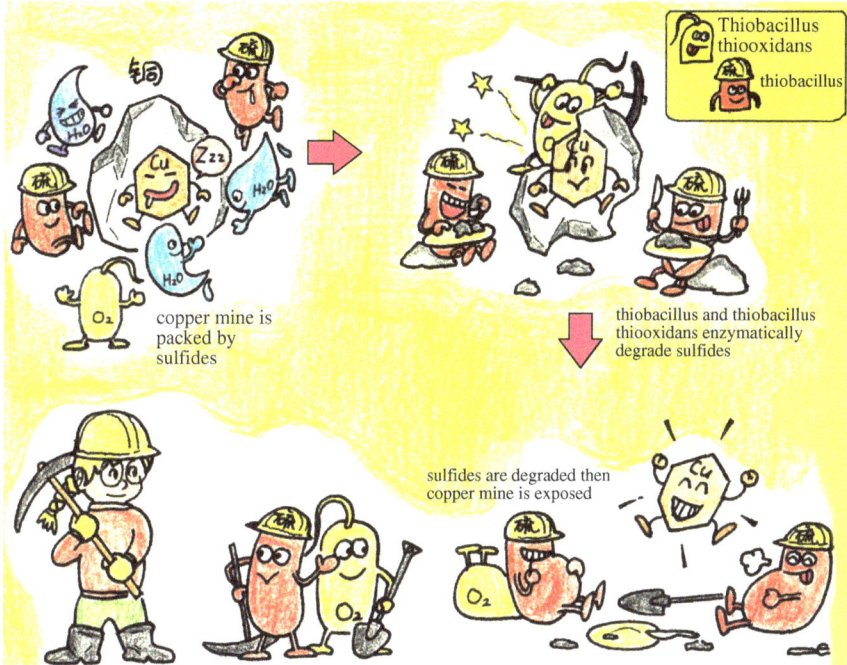

Figure 2-7: Bacteria and metal smelting.

and taste. It is made mainly produced from rice, millet, or sorghum, with rice pieces, corns, dried sweet potatoes, and potatoes as raw material. Raw material will go through steaming, gelatinization, liquefaction, and saccharification. To convert starch into sugar, enzymes are used to ferment the material to ethanol, which is then oxidized by *Acetobacter* to acetic acid.

Dairy products

Fermented dairy products have special flavors. They are produced with good quality raw milk, which is sterilized, inoculated with certain bacteria, and then fermented. They usually taste good, have high nutritious values, and serve certain health effects. The most common fermented dairy products include yogurt, cheese, sour cream, kumis, and many others. Bacteria that play a main role in this process include *Lactobacillus casei, Bifidobacterium, Lactobacillus bulgaricus, Lactobacillus acidophilus, Lactobacillus plantarum, Lactobacillus,* and *Streptococcus thermophilus.*

Brewery

Brewery in ancient times relies on koji. What is koji? Koji is a compound of many microorganisms including yeasts, molds, and bacteria. Under the reaction of these microorganisms, cereals are made into wine through saccharification, fermentation, and other processes. Bacteria play a unique role in brewery. During fermentation, introducing the appropriate bacteria can solve the problem of lacking of aftertaste in liquor; metabolites of bacteria can improve the aroma and flavor of liquor, Huangjiu, and wine. The most commonly used bacteria in brewery include *Lactobacillus, Acetobacter,* and *Clostridium butyricum.*

  Lactobacillus can ferment carbohydrates into lactic acid, and lactic acid produces ethyl lactate through esterification. Ethyl lactate provides liquor a unique aroma. However, excessive lactic acid will increase the acidity of the fermented grains, thus influencing the quantity and quality of yielded liquor. Too much ethyl lactate makes the liquor less sharp. Acetic acid produced by *Acetobacter* is one of the main components of liquor aroma; however, too much acetic acid will provide the liquor a pungent acid smell. *Clostridium butyricum* produce butyric acid through fermentation, which provides a rich and lasting taste to the liquor.

  In addition, *Pediococcus halophilus* participate in brewing soy sauce; lactobacilli and *Leuconostoc* participate in kimchi production. Bacteria provide many types of tasty food for humans (Fig. 2-8).

Working magic in modern agriculture

Nonpolluting biological pesticides

Biological pesticide refers to animals, plants, and microorganisms that can prevent and cure plant diseases. Back to early 20th century, a flour factory in Thuringia, Germany, encountered something

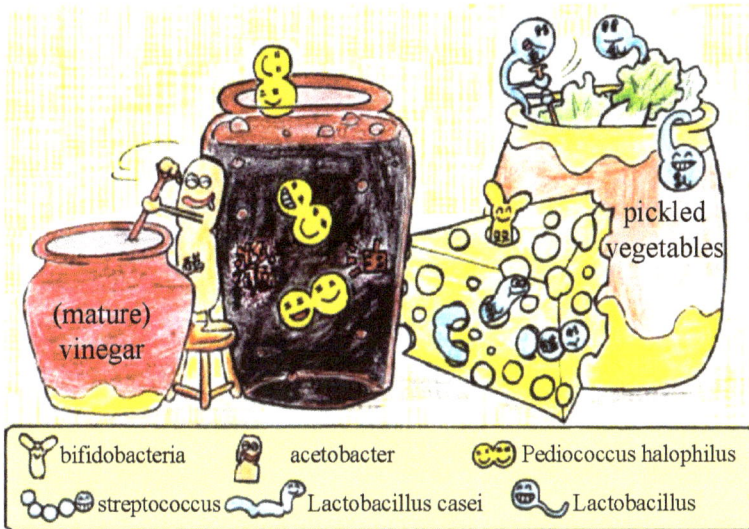

Figure 2-8: Role of bacteria in food.

strange. Larvae of Mediterranean flour moth that used to fly around in the warehouse suddenly died in large numbers. Biologist Berliner was very interested in the phenomenon. In 1911, he isolated a rod-shaped bacterium from the dead insects. This bacterium was named *Bacillus thuringiensis* in 1915. Berliner brushed the bacterium on leaves. When the moth larvae ate these leaves, they were first agitated, and then they died in 2 days. Later, Berliner discovered that shortly after the spores of this bacterium are formed, they form square or diamond-shaped crystals, known as parasporal crystals. Unfortunately, this discovery had not been taken seriously. Between 1920 and 1950, many used *Bacillus thuringiensis* to conduct field trials of pest control. Until 1956, a biologist named Hannah proved that the parasporal crystal was the real reason for the death of moth larvae (Fig. 2-9).

Bacillus thuringiensis can kill more than 130 types of pests including Lepidoptera, Hymenoptera, and Orthoptera. When *Bacillus thuringiensis* enters the digestive system of the pest, parasporal crystal present in the biological pesticide is digested and activated by alkaline intestinal fluid. It turns toxic, and the pest dies of intoxication.

Figure 2-9: Biological pesticide *Bacillus thuringiensis*.

However, intestinal fluid of humans and animals is acid, thus it cannot dissolve this crystal protein. Therefore, *Bacillus thuringiensis* is harmless to humans and animals.

Other biological pesticides include *Bacillus scarab* and *Bacillus sphaericus*. When ingested by scarab larvae, *Bacillus scarab* grows in the intestines and becomes active nutrient bodies. These bodies cross the intestinal walls, reproduce inside the body, damage organs, and finally kill the scarab. Parasporal crystals in *Bacillus sphaericus* are particularly effective to mosquito larvae.

Some bacteria can also be used as herbicides. *Puccinia chondrillina*, for example, can control gum succory, the main weed in wheat fields; *Xanthomonas campestris* can remove weeds from lawns. This new type of herbicide is safe and convenient and will not cause pollution. It represents the development trend in herbicides.

Apart from bacteria that are used as biological pesticides, other microorganisms that are used as pesticides include bassiana, *Metarhizium, Paecilomyces lilacinus* (control of plant nematodes), virulence Entomophthorales (control of aphids), *Verticillium* fungus (control of whitefly), *Gliocladium* (control of carrot blight), *Bacillus subtilis*, *Pseudomonas*, and *Bacillus cereus*. Some viruses are also used in pest control, for example, nuclear polyhedrosis virus (NPV), granulovirus (GV), cytoplasmic polyhedrosis virus (CPV), and insect poxvirus.

Biochemical pesticides with low toxicity and high effectiveness

Biochemical pesticides are obtained from metabolites of animals, plants, and microorganisms. At low concentrations, these can prevent and cure plant diseases. Biochemical pesticides derived from animals include insect juvenile hormone, sex pheromones, and bee venom. Those derived from plants are mainly phytotoxins, namely substances produced by the plant that intoxicate and kill the insect, or introduce specific reactions to the insect (e.g., aversion to food, inhibition of growth, repellent, and inhibition of spawning). All antibiotics and microbial toxins produced by microorganisms belong to biochemical pesticides with microorganic origin. Agricultural antibiotics (namely biochemical pesticides from microorganisms) are secondary metabolites produced by microorganism fermentation. At low concentrations, they can detain or eradicate a disease, pest, or weed for the crop, and control crop growth and development. Japan is the leading country of agricultural antibiotics in the world. It has developed avermectin, kasugamycin, blasticidin, polyoxin, validamycin, polynactin, and many others. Among them, avermectin is by far the most effective pesticide antibiotics. It is a macrolide antibiotics produced from *Streptomyces avidinii*. It is widely used and is highly effective with low toxicity. Validamycin can cure rice with sheath blight. When validamycin comes in contact with the hyphae of the pathogens, it can be immediately absorbed and transferred into the cell. It disturbs and inhibits the normal growth of the bacterium and deprives it of ability to invade, thus curing the crop (see Fig. 2-10).

In addition, herbimycin is a herbicide produced from bacteria. It only deters growth of weeds but has no effect on green plants such

Figure 2-10:   Validamycin curing sheath blight.

as rice, which is due to the differences in the growth of rice and weeds. Herbimycin can exactly deter the part of growth of weed that is different from rice; therefore, it does not affect the rice. Another bacteria-produced herbicide is dipropylene phosphorus. It is highly effective on more than 100 types of annual and perennial weeds.

Compared to chemical pesticides with similar functions, agricultural antibiotics produced from bacteria are often less toxic and do not leave remains.

Bio-fertilizers

Broadly speaking, bio-fertilizers include plant, animal, and microbial fertilizers. In the narrow sense, bio-fertilizers refer only to microbial fertilizers. For a plant to grow, it must absorb from its environment various nutrients such as carbon, hydrogen, oxygen, nitrogen, phosphorus, potassium, sulfur, calcium, magnesium, iron, copper, manganese, zinc, boron, and molybdenum. Plants require a considerable amount of the first 10 nutrients, which are known as macronutrients; they need the remaining nutrients in smaller amounts, which are known as micronutrients. Among them, carbon, hydrogen, and oxygen can be obtained from carbon dioxide in the air and water in soil; apart from certain areas that lack some micronutrients, most soil contain sufficient supply. But soil supply of nitrogen, phosphorus, and potassium is insufficient, yet plants have large requirement of these elements. Therefore, nitrogen, phosphorus, and potassium are considered the three fundamental factors for plant growth.

In general, green plants can neither obtain the required nitrogen directly from air nor absorb complex nitrogen-containing organic molecules from the soil. Similarly, they cannot directly absorb phosphorus and potassium from soil. Therefore, where do these three nutrients come from? Why can some plants flourish without fertilization? The credit indeed goes to bacteria.

Diazotroph

There is a kind of bacteria called diazotroph that can fix nitrogen in air to the root of plants for them to absorb. Among these bacteria, rhizobia are most commonly used to increase soil fertility. When rhizobia enter plants, they form small lumps at plants' roots. These lumps will fixate nitrogen floating in the air and transfer it into usable nitrogen for the plants. One can say these small lumps are small "fertilizer factories" established at plant roots (Fig. 2-11). Most legumes have these lumps on the root. Therefore, even if planted in relatively poor soil, legumes can survive on little fertilizer. Furthermore, soil on which legume has been planted turns out to be quite fertile even though little fertilizer has been applied. Thus, people in rural areas often grow different plants at different times on the same lot of soil.

Phosphate-solubilizing bacteria

"Phosphate-solubilizing bacteria" is the overall term for bacteria in the soil that have a strong ability to dissolve phosphate compounds.

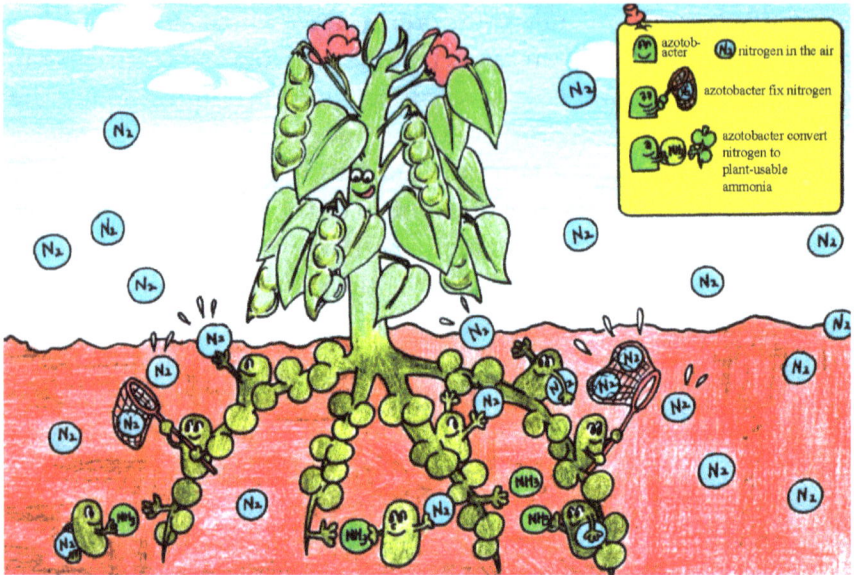

Figure 2-11:   Mutually beneficial relationship between rhizobia and legumes.

By producing various enzymes and acids, phosphate-solubilizing bacteria transfer phosphide in the soil into soluble phosphide that can be used by plants. These are of two main types. One is organic phosphate bacteria, whose main function is to decompose organic phosphorus compounds such as nucleic acids and phospholipids; the other type is inorganic phosphate bacteria, whose main role is to decompose inorganic phosphorus compounds such as calcium phosphate and ash stone. In agriculture, phosphate-solubilizing *Bacillus megaterium* (also known as *megaterium* bacteria) is often used. Other bacteria that are used in fertilizers include some bacilli, *Achromobacter* and *Pseudomonas*. It has been proved in practice that phosphate-solubilizing bacteria can increase the production of crops such as wheat, sweet potatoes, soybeans, and rice, as well as fruit trees such as apples and peaches.

Potassium bacteria

"Potassium bacteria" is the general term for all bacteria in soy that have strong ability to dissolve silicate compounds. On the one hand, these bacteria produce organic acids through metabolism, which dissolve potassium and phosphorus from feldspar, mica, apatite, phosphate powder, and other minerals. Potassium and phosphorus are used by the bacteria; after the bacteria die, potassium in their cells will be absorbed by crops; on the other hand, potassium bacteria can also produce hormones, amino acids, and polysaccharides, which promote the growth of crops. In the meantime, potassium bacteria reproduce in the soil and detain the growth of other pathogens. All these are beneficial for the growth of crops, the increase of production, and the improvement of the crop quality.

When scientists discovered the functions of these bacteria, they isolated them from the soil and cultivated them on a large scale. In the end, these bacteria are used as fertilizers in agriculture, as shown in Fig. 2-12. Some bacteria can decompose animal and plant remains in the soil to provide nutrients for crops. This is the main source for the three crucial elements necessary for crop growth: nitrogen, phosphorus, and potassium. Some bacteria grow within the rhizosphere of the crop. They release substances that stimulate and control crop growth, promote budding, and reduce pest diseases.

After the earlier discussions, you probably already know why some plants can flourish without fertilization and also how bacteria can be used as fertilizers.

Contribution to environmental management and protection

Human societies attained great achievements in the 20th century. World economy has seen unprecedented development. Behind the great developments, however, hide potential catastrophes. Chemical wastes accumulate on hills; toxic substances leak out in large quantities, and many rivers and coasts are seriously polluted. Total suspended particulates exist in large amounts in the air. Around the world, ecology is deteriorating; we are concerned about the future of the Earth.

Activated sludge: a large army combating wastewater

Most sewage and industrial wastewater contains considerable amount of toxic and harmful organic substances. Should the sewage

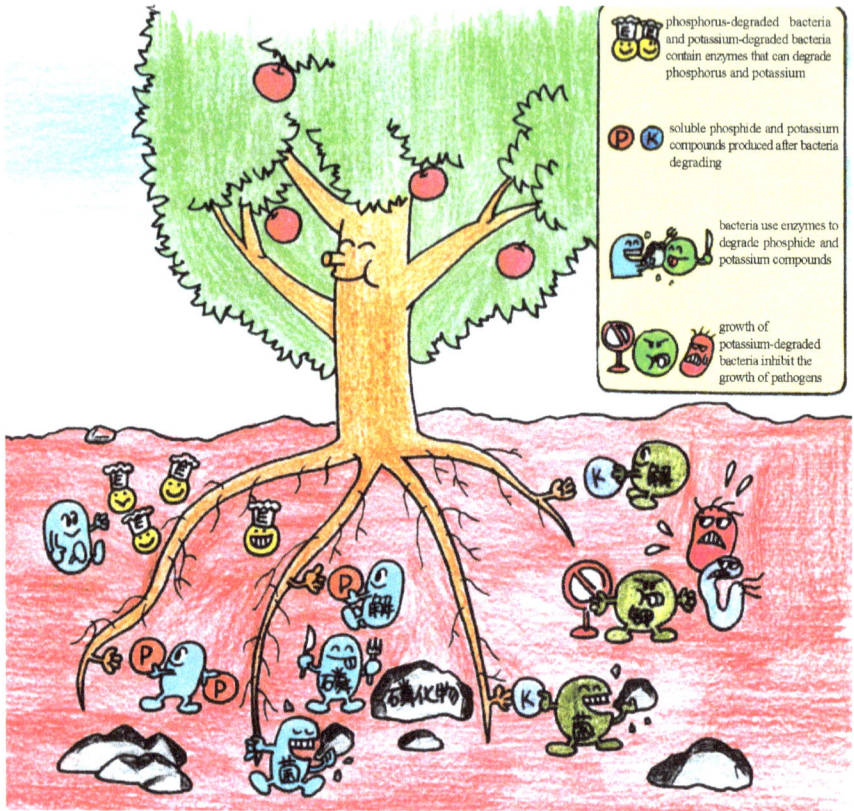

Figure 2-12: Phosphate-solubilizing bacteria and potassium bacteria as bacteria fertilizers.

be discharged directly without treatment, it would contaminate soil and water and enter human body via the food chain, thus threatening human health. Therefore, almost all sewage and industrial wastewater are processed with activated sludge before being discharged into the environment. The so-called activated sludge consists of cotton-like substances formed by various microorganisms and protozoa living together, combined with some nutrients and floating debris from wastewater. Microorganisms in the activated sludge work together. They play to their strengths to absorb pollutants in wastewater into their cells. The toxic and harmful substances in wastewater are therefore transformed into a component of the bacteria, or they are dissolved into carbon dioxide and water. Meanwhile, activated sludge has strong adsorption capacities. It can absorb many pollutants, thus fulfilling the goal of wastewater treatment and purification. The major members in activated sludge are bacteria, including *Alcaligenes, Microbacterium, Bacillus, Pseudomonas,* and others. Figure 2-13 shows the common process of wastewater treatment.

As a wastewater treatment method, activated sludge has more than 90 years of history. It can be traced back to 1912, when Clark and Gage from Britain found that when exposed in air for long periods, wastewater would produce sludge. Meanwhile, quality of water was significantly improved. Later, Arden and Lockett further studied

Figure 2-13: Activated sludge treating wastewater.

this phenomenon. Aeration tests were conducted in bottles. Every day when the experiment was over, the bottles were emptied; it started over again the next day. Arden and Lockgtt discovered accidentally that if the bottle was not completely cleaned and there were sludge attached to bottle wall, water treatment could even achieve a better result. Realizing the importance of sludge remaining in the bottle, they named it activated sludge. Later, before the experiment was over each day, they put the aerated water still for precipitation, tossing away only the purified water above and leaving the sludge in the bottle. The sludge was then saved for use the second day. This greatly reduced the time for water treatment. Observing the brown, flocculent sludge under microscopes, one could view many bacteria, fungi, protozoa, and metazoan. They made up a unique ecosystem. Precisely, because these microorganisms (bacteria mainly) feed on organic substances present in wastewater to grow and reproduce, the amount of organic content in wastewater can be reduced. Activated sludge includes aerobic activated sludge and anaerobic granular sludge. The history of activated sludge marks 1914 as the founding year of this method. In 1921, the first wastewater treatment plant using activated sludge was established in Shanghai, China.

Bacteria specifically used for green plastic bags

Plastic bags and plastic boxes made from petroleum bring convenience to our lives; however, because they have highly stable chemical properties, they are unlikely to be degraded under natural conditions, thus causing "white pollution". Debris of plastic films in farmland can stay as it was after buried in soil for 50–100 years. This will hamper crop growth and reduce production.

How do we deal with "white pollution"? How do we avoid it? The most effective way is to find alternatives to plastics that can be naturally degraded by microorganisms. Apart from paper products or products of natural animal and plant fibers, there is also a bacterium called *Pseudomonas cepacia* that can produce polyhydroxybutyrate (PHB) similar to plastics [also known as biodegradable plastics; it is one type of polyhydroxyalkanoates (PHAs)]. Accumulated PHB can weigh up to 60% of the cell. This polyester can be completely

Figure 2-14: PHA production and degradation process.

degraded in nature, and its degradation products serve as fertilizers that can improve soil structure. In addition, as this type of plastics has other features such as anti-UV, nontoxic, noninflammatory, transparent, and easy to color, it is also widely applied for medical uses. For example, new orthopedic devices such as artificial joints fabricated from degradable plastics have a great advantage: once it is implanted in the body, it automatically decomposes in a certain period and is absorbed by tissues; by then, the bones are already healed. A second operation to take out the device is no longer necessary. This also relieves patients of extra pain and trouble. Figure 2-14 shows white pollution caused by traditional plastic products, as well as the degradation process of degradable plastics made with bacteria.

Bacteria specializing in soil remediation

Currently, bioremediation is the main technology of remediating polluted soil. Bioremediation technology is the process where microorganisms "eat" organic pollutants in the soil and convert them into carbon dioxide and water or other harmless substances.

As natural bioremediation process is slow, industrialized bioremediation is introduced. It is an artificially promoted remediation. The degradation capacity of microorganisms can eliminate petroleum or other toxic and harmful pollutants from soil. Degradation can be realized through modification of physical and chemical conditions of soil (temperature, humidity, pH, aeration, and nutritional additives); it can also be achieved by inoculating specially engineered microorganisms that can increase the degradation rate. Remediation of organic-polluted soil with microorganisms is based on degradation and conversion of pollutants. It includes aerobic and anaerobic processes. In aerobic processes, microorganisms can degrade organic pollutants in soil and convert them into carbon dioxide and water; in anaerobic processes, the main products are organic acids and others (methane and hydrogen, for example). Two common techniques used in biotreatment of polluted soil are in situ and ex situ remediation.

Scientists planted some special bacteria in the soil polluted by oil leaked from a nearby oil storage tank; they also put some nutrients in the soil at the same time. It turned out that in less than 60 days, these bacteria consumed 80% of the oil in the soil. At the beginning of the experiment, consumption of oil by bacteria was not satisfactory; later, scientists had to implant more and more bacteria and restrain their activities within the polluted area in the soil. Scientists believe that, to accelerate bacteria's degradation of oil, it is essential to "concentrate their force to defeat the enemy". Scientists emphasize that, compared to burning, cleaning, and other methods of freeing soil from oil pollution, the cost of biological methods is lower. Researchers pointed out that when an oil tanker sank and its surrounding sea was polluted, people can first apply physical methods to remove the leaked oil — this is the more effective method at this stage. However, when oil floats to land and contaminates the soil, people should use biological methods, which are more suitable for this situation. Soil contamination by oil is regional; this makes it easier to apply biological method for oil removal.

During the Gulf war, large amounts of oil wells were bombed in Kuwait, causing serious soil pollution. To solve the problem, special-

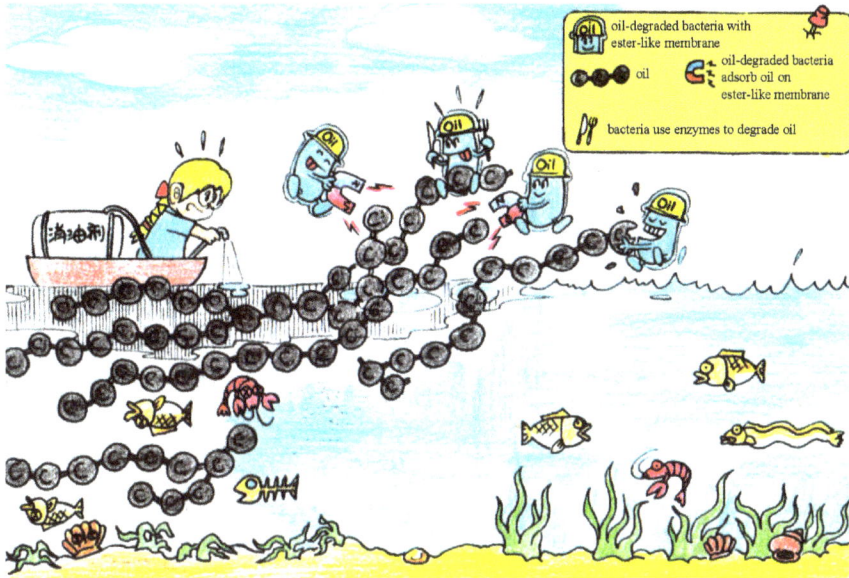

Figure 2-15: Bacteria treating oil pollution on the surface of water.

ists discovered a new type of bacterium that feeds on petroleum. This bacterium has a film made of ester at its surface, which makes absorbing oil easier, thus restoring the soil to its original state. Figure 2-15 shows how bacteria treat oil floating on water. Figure 2-16 shows bioremediation process of soil after contaminated with petroleum.

A methane-eating bacterium is cultured, selected, and mutated by American scientists to acquire trichloroethylene bacteria that can degrade important components in industrial wastes. On the basis of this, a large range of other microorganisms are cultured for environmental protection, for example, mercury-resistant bacteria that can degrade highly toxic compounds. These bacteria enter the soil and battle together, so that contaminated soil is clean again.

Scientists have also discovered bacteria that can remove harmful metal: some can convert metal and radioactive substances (arsenic, mercury, lead, tin, uranium, etc.); some can accumulate metal in their body, which reduces pollution and maintains ecological equi-

soil polluted by oil before repair

soil deeply polluted by long term soaking of oil

soil during repair

soil after repair

Figure 2-16: Bioremediation process of petroleum-contaminated soil. (From Introduction to Bioindustry by the same author.)

librium. These special bacteria cannot be poisoned by harmful substance; on the contrary, they devour these substances. This makes them the guardians of human life and health.

New energy in the post-oil age

Lack of new energy and environmental pollution is the main problem faced by social and economic development. The energy issue is becoming increasingly prominent. The petroleum, coals, and gas we use are nonrenewable resources, and these will be used up one day; their use also causes a series of problems: petroleum and coal cause serious pollution such as acid rain and dust in the air. Although new energies such as nuclear, solar, and wind energy are gradually being used, there are still many restrictions. Nuclear power, for example, is strongly rejected by many environmental protectionists around the world for its potential dangers. By far, using solar energy to produce electricity is very costly and is considerably influenced by weather conditions. Wind power requires high investment and is subject to regional resources. Although China encourages water power, its impact on ecology cannot be neglected. Thus, bioenergy as a renewable and low-pollution green energy is attracting increasing global attention. Bioenergy refers to energy substances generated through treatment of abandoned or cheap biological material, such as straw, cassava, sugarcane, and all types of animal wastes, by microorganisms (mainly bacteria). For example, cyanobacteria and *Pseudomonas* can produce H_2 and biodiesel from straw, as shown in Fig. 2-17. As the material used by bacteria is renewable, bioenergy is also known as "renewable energy". Currently, there are four main types of bioenergy: marsh gas, biohydrogen products, ethanol fuel, and biodiesel.

Figure 2-17: Bacteria and bioenergy.

Cheap marsh gas

Marsh gas is a compound gas. As it was first discovered in marshes, it was thus named marsh gas. Marsh gas fermentation is also known as anaerobic digestion. Human and animal feces, straws, wastewater, and other organic matters are sealed in a digester. Under anaerobic conditions (no oxygen), these materials are fermented, degraded, and converted by microorganisms, and marsh gas is produced.

In hot summers, one can often see air bubbles of all sizes popping up from marshes, cesspools, and septic tanks. If the gas is collected in a glass bottle and lighted, there will be light-blue flames at the top of the bottle. This is marsh gas. It is a compound of many gases, including methane, carbon dioxide, nitrogen, oxygen, hydrogen, hydrogen sulfide, carbon monoxide, and steam. Among them, methane makes up 55%–70%; carbon dioxide is 30%–45%. Marsh gas can be used in power generation and lighting. It can also be used as domestic and industrial fuel.

Swamp gas is an ignitable gas. It is produced when bacteria and other microorganisms ferment and degrade straw, grass, and feces under anaerobic conditions. There are many bacteria that can produce marsh gas. In soil, lakes and marshes, pond sludge, gastrointestinal tracts of cows and sheep, feces of horse and cattle, and garbage piles, there are always a considerable amount of methane bacteria. Figure 2-18 shows straw, animal feces, and other wastes. It also shows how they are converted into marsh gas.

Biohydrogen products

Hydrogen is another important source of energy. As hydrogen bonds possess high bond energy, burning 1g of hydrogen can release 142 kJ of heat. This is three times more efficient than gasoline. In addition, hydrogen is light and easily portable compared to gasoline, natural gas, and kerosene. Therefore, it is a suitable fuel for high-speed flight vehicles used in aerospace and aviation. Hydrogen reaction with oxygen can produce flames of up to 2500 °C. Thus, hydrogen is also an excellent material for cutting or welding metal. Combustion of

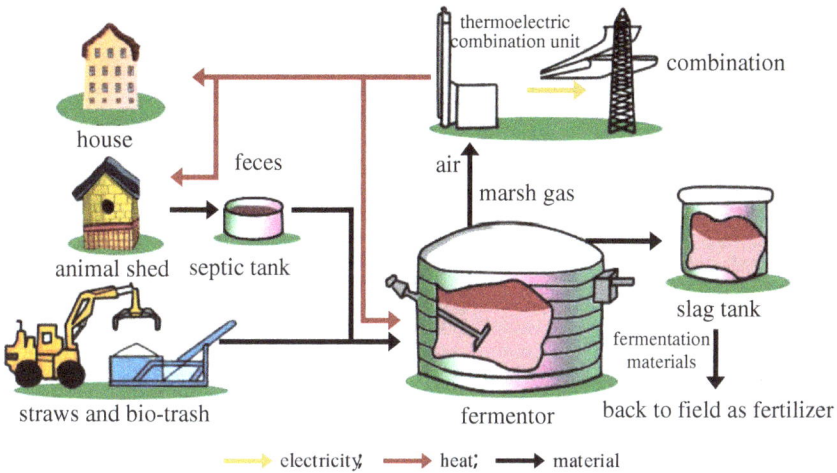

Figure 2-18: Conversion process of marsh gas from wastes such as plant straws, animal feces, and biological trashes. (Adapted from Introduction to Bioindustry by the same editor.)

hydrogen produces only steam. If hydrogen is used, the problem of automobile exhaust pollution will also be eliminated. Therefore, hydrogen is the ideal clean energy.

Hydrogen exists everywhere on the earth. It is in water, carbohydrates, and hydrocarbons. But there is not a considerable amount of free hydrogen in nature. It can only be obtained by decomposing hydro-compounds. Finding a way to acquire hydrogen from water, carbohydrates, and hydrocarbons has been a problem worldwide. Since Nakamura first discovered that microorganisms produce hydrogen in 1937, bio-hydro production has been the first prime research objectivity of hydrogen production. Biohydrogen production is a technique that uses certain microorganisms to obtain hydrogen from water and other organisms under normal temperature and pressure, and in a low-energy-consuming, environmental-friendly way. By far, more than 20 types of bacteria and eukaryotic algae are proved to be able to produce hydrogen. On the basis of their different methods of producing hydrogen, these microorganisms can be divided into three groups: cyanobacteria and microalgae, which produce hydrogen through water pyrolysis in light; purple photosynthetic bacteria, which produce hydrogen through fermentation in light; and anaerobic or facultative anaerobic microorganisms, which produce hydrogen through fermentation without light.

Ethanol fuel

Ethanol fuel is also known as alcohol fuel. When bacterium ferments plants, it produces ethanol with high moisture. After further dehydration, a suitable amount of gasoline is combined with ethanol to form ethanol fuel. Ethanol fuel can not only reduce the use of gasoline but also increase the gasoline efficiency, thus avoiding efficiency-promoting additives that are harmful to the environment, such as tetraethyl lead.

Raw material for bacteria production of ethanol fuel includes corn, sorghum, wheat, barley, sugarcane, sugar beet, potatoes, as well as municipal waste, bagasse, small tree trunks, wood pieces, and other fibrous materials. Producing alcohol from grain is relatively simple. After grounding the grain, amylase and glucoamylase are added to hydrolyze starch into glucose. Then, bacterial fermentation is used to convert glucose into alcohol. But today, there is food shortage in the world. Even if the remaining grains are all used for alcohol production, they can only replace a small portion of petroleum. Therefore, scientists are researching on using straws as an alternative to grains. Cellulose and hemicellulose in straws are hydrolyzed into glucose and xylose; then, bacterial fermentation turns them into alcohol. Enzymes produced by bacteria convert big molecules into small molecules, as shown in Fig. 2-19.

Biodiesel

Diesel is the main source of fuel for many large vehicles such as trucks, diesel locomotives, and electric generators. There is a large demand of diesel. Incomplete combustion of conventional diesel will emit large amounts of CO_2, which will cause acid rains and greenhouse effects. It is a major cause of environmental pollution.

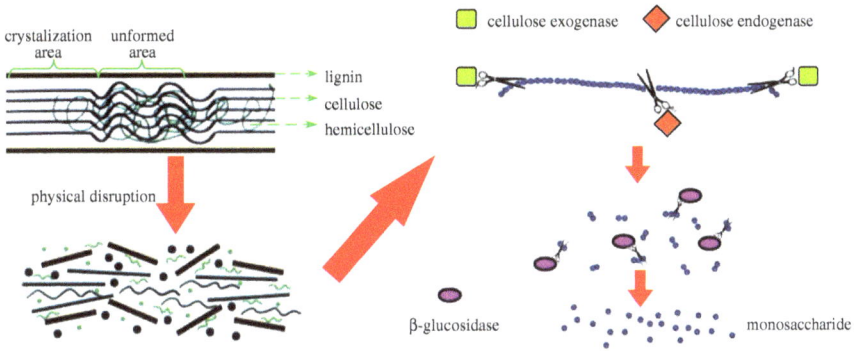

Figure 2-19: Enzymes produced by bacteria convert big molecules into small molecules. (From Introduction to Bioindustry by the same author.)

To solve the problem of exhaust pollution and to alleviate the pressure of deteriorating environment, countries around the world are investing in new energies. Replacing conventional diesel with biodiesel is one of the effective solutions.

Biodiesel comprises fatty acid methyl ester or fatty acid ethyl ester compounds obtained through the ester-exchange interaction between long-chain saturated and/or unsaturated fatty acid and alcohol (methanol or ethanol). Because it uses plant fat, animal fat, and other biological resources as the raw material, it is called "biodiesel". Compared to conventional fossil energy, biodiesel has the advantages of being nontoxic, biodegradable, low in sulfur and aromatic content, and good lubricity. It can blend with fossil diesel at any ratio. It also protects the engines. Producing biodiesel with waste oil is an act of converting waste into valuable resources. It not only reduces air pollution caused by diesel vehicles but also prevents oil entering market such as cooking oil after being processed, which will be harmful to health.

Recently, British scientists obtained a type of *Escherichia coli* that can produce bioenergy equivalent to conventional gasoline. The principle is to assemble enzymes from *Photobacterium, Candida albicans, Bacillus subtilis,* and camphor into *Escherichia coli* and to convert normal *Escherichia coli* fatty acid into saturated alkane (hydrocarbons), so as to produce large quantities of gasoline and diesel. However, this process is by far low in efficiency. Researchers have not been able to apply it to industry. More optimistically, though, scientists predict that in the coming 10–15 years, oil production can be achieved using microorganisms.

"Waste oil" reused and put back on the table is abominable. Currently, waste oil can be converted to biodiesel through pretreatment, esterification, transesterification, alcohol concentration, and distillation. In addition, research by British scientists reveals that waste cooking oil can be an efficient raw material to synthesize biodegradable plastic with the help of a specific bacterium. Once large-scale production is achieved, it can reduce environmental pollution as well as can provide high-quality plastics for medical implants.

Apart from the four types discussed earlier, microbial fuel cell is also a type of renewable energy. It is a device that uses microorganisms to convert chemical energy in organisms directly into electrical energy. Its basic working principle is as follows: first, in anaerobic environment of anode chambers, organisms release electrons and protons under microorganic degradation; then, electrons are conveyed through suitable protons media between anode and biological components; furthermore, electrons are delivered to the cathode with outer circuit to form a current; meanwhile, protons are transported to cathode through proton exchange membrane; finally, an oxidant (typically oxygen) gains electrons at the cathode and combines with protons to form water, as shown in Fig. 2-20. Professor Derek Lovley from University of Massachusetts Amherst, the United States, has been conducting related research in his laboratory. His work is chosen by *Time* magazine as one of the 50 most important inventions in 2009.

Bacteria friends awaiting discovery and invention

Bacteria used in powder metallurgy

Traditional brick production uses fire and produces large amounts of carbon dioxide. Today, scientists plan to use bacteria for the consolidation process to make brick production environmentally friendly. To explain the process briefly, first, certain bacterial broth is put into the sand; then, after a week, the sand gets converted to

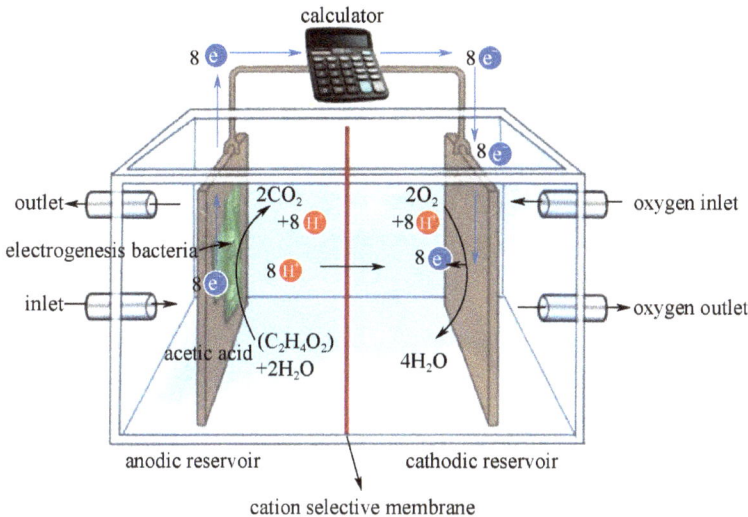

Figure 2-20: Basic principle of microbial fuel cell. (From Introduction to Bioindustry by the same author.)

"solid" stones by these bacteria. Farewell to fire! Moreover, raw material for such bacteria brick is sand instead of clay! Basically, it has the same function as fire-produced bricks, because the special bacteria can secrete powerful coagulants. As long as there are no psychological barriers bothering you — for example, that the bricks are full of dead bacteria — then this brick can build a better future. However, further scientific research is required to find whether bacteria brick can be used in constructions. Figure 2-21 shows the production process of bacteria bricks. Inside the little bottle is the broth of this particular type of bacteria.

Bacteria feeding on Asadin

Talking about arsenic, most people will think of its highly toxic compound, Asadin. The American scientist Felisa Wolfe-Simon recently discovered a unique bacterium that uses arsenic instead of phosphorus to build biomolecules.

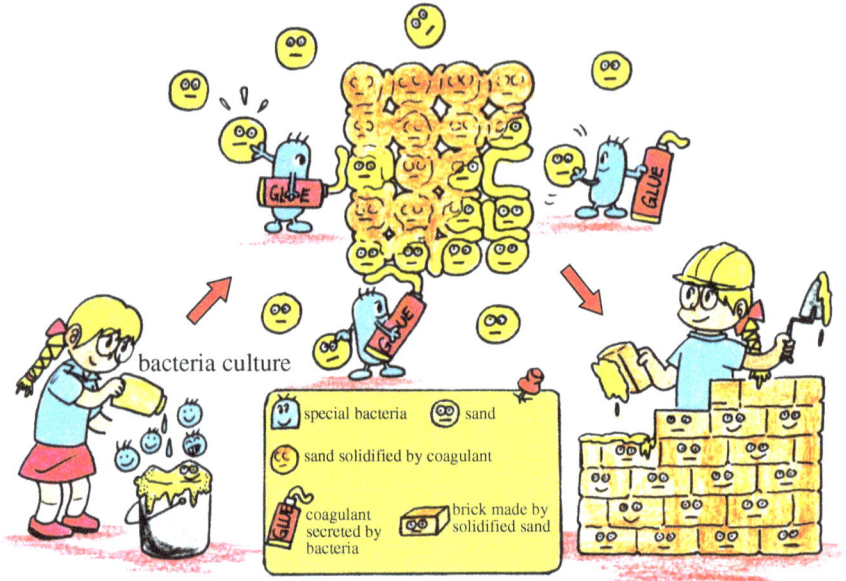

Figure 2-21:   Production process of bacteria brick.

According to the existing scientific knowledge, carbon, hydrogen, oxygen, nitrogen, and sulfur are the six fundamental elements that form life on the earth. Phosphorus plays extremely important role in the cell, including preservation of genetic material and participation in all life activities. However, Felisa Wolfe-Simon found in Mono Lake in Eastern Carolina a type of bacteria that uses arsenic to replace phosphorus as the necessary substance for its maintenance. As the surrounding areas of Mono Lake are rich in arsenic, there is high concentration of arsenic in the lake. Scientists collected sludge from bottom of the lake and cultured it with high-arsenic, low-phosphorus artificial salt. After a series of dilution processes, most phosphorus in the solution was removed and replaced with arsenic. The result revealed that one type of bacteria grew significantly better than others. Isolated culture confirms that this type of bacteria can grow in arsenic solutions 60% as fast as in phosphorus solutions. When there is neither arsenic nor phosphorus in the environment, it cannot grow.

This is the first discovery that fundamental elements of life can be replaced by other elements. This discovery will bring significant change to human understanding of life and open up new possibilities to look for life in extreme or extraterrestrial environments.

Engineered bacteria: "cell factory" at our service

With developments in life sciences and biotechnology, scientists can artificially process and transform bacteria. Bacteria are thus turned into cell factories that can produce various products at our demand. The basic principle of constructing such cell factories is to obtain useful life elements from bacteria, animals, and plants, and to plant these elements into a certain bacterium like a circuit. This "synthetic bacteria" can achieve different biological functions based on our expectations and produce substances that we need. In 2003, the University of Massachusetts set up a standard registry of biological parts. So far, it has collected around 3200 standardized biological parts, which are accessible to scientists around the world. On the basis of these parts, specific biological systems can be assembled. Mostly, they are used to assemble "synthetic bacteria". Preparation and production of synthetic microorganisms is called "synthetic biology".

Currently, synthetic biology is being used in industries. Two American entrepreneurs are using synthetic bacteria to produce biofuels. Recently, according to pathology of *Pseudomonas aeruginosa*, a pathogen that seriously threatens human life, scientists purposefully transformed *Escherichia coli* and obtained a "synthetic *Escherichia coli*" that can effectively inhibit the growth and reproduction of this pathogen (see Fig. 2-22). This provides new possibilities to drug development and clinical treatment.

Bacteria that conquer tumors

More than 100 years ago, William B. Coley discovered that when patients of sarcoma suffer from acute streptococcal infection, their immune systems get activated, suppressing and eliminating the tumors. Later, it was discovered that other bacteria could also pen-

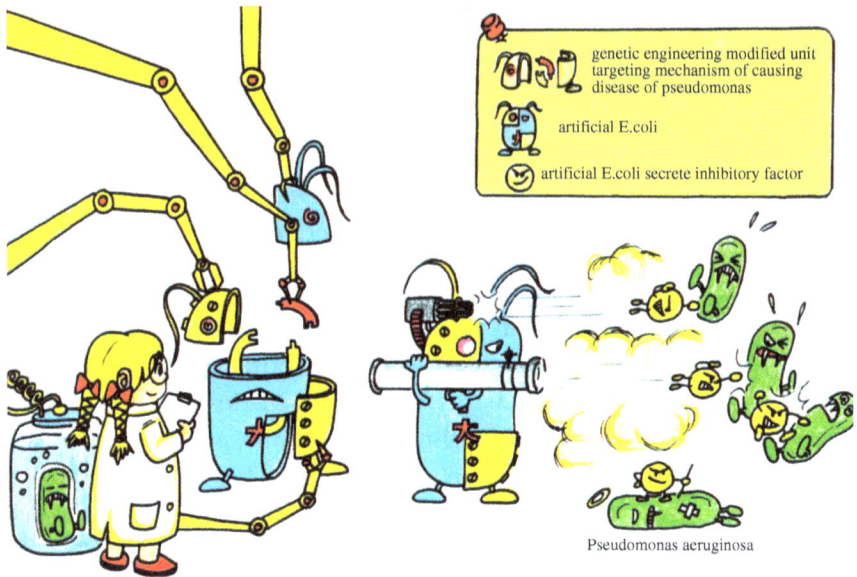

Figure 2-22: Synthetic bacteria.

etrate the body, accumulating preferentially in tumors. Bacteria, especially anaerobic and facultative anaerobic bacteria, have this special feature. It is possible to create effective tools in combination with current treatments for curing tumor more effectively. It is also possible to add antitumor properties to bacteria through genetic engineering, thus increasing their effectiveness and decreasing their toxicity. Latest research shows that certain proteins produced by bacteria can significantly increase the effectiveness of other treatments. This is a new perspective to tumor treatment: to use not the whole living bacteria, but only its protein. This can avoid harmful side effects of living bacteria. Obligate anaerobes do not require oxygen to grow; they cannot tolerate oxygen. Scientists discovered that only Novi Clostridium ATCC19402, Delhi Clostridium ATCC9714, can be widely planted in necrotic areas of tumor. When mice were injected with spores of these bacteria, there was no significant toxicity; more importantly, the growth of bacteria was restrained inside the tumor. They were not detected in other organs such as the liver, spleen, kidneys, lungs, or the brain. However,

16–18 h after the injecting spores, all mice died of lethal toxins from bacteria germination. Later, toxin-free *Clostridium novyi* was acquired by removing toxin-encoding genes; when spores were injected in veins, modified bacteria colonized tumor as previous bacteria did and effectively increased the necrotic area.

Furthermore, as these anaerobic bacteria colonize tumor cells easily, scientists combined this special feature with other antitumor drugs. This treatment attacks tumor cells from both interior and exterior and forces tumor cells to "surrender". Some scientists discovered that these anaerobic bacteria have the feature of "targeted" tumor cells colonization. They can also produce "liposome enzyme". Therefore, when these bacteria are combined with "liposomal drug" to treat mice with tumor, it increases the concentration of drug in tumor cells, thus killing tumor cells more effectively. The benefits of bacteria are in every aspects of our life (Fig. 2-23). Some bacteria are indeed our beloved friends.

Figure 2-23: Benefits of bacteria.

Bacteria Are Our Detestable Enemy

Within the bacteria family, many bacteria can bring us various benefits, but there are a small number of bacteria that are our enemies. In the long human history, many disastrous infectious diseases were caused by bacteria. Many among them still threaten our lives. These bacteria that cause disease to human bodies are called pathogens. Generally, among all types of bacteria and hosts (where bacteria live) in human bodies and among bacteria themselves, there is a good balance due to control factors such as competition for nutrition and metabolite. Under certain conditions, the balance is broken. Formally, nonpathogenic normal bacteria can become pathogens. These are called opportunistic pathogenic bacteria, which are also known as opportunistic pathogens. For normal bacteria to turn into pathogens, three conditions need to be fulfilled: first, their place of settlement changes. Some bacteria leave their normal habitation and move into other places. Away from previous restrains, they grow and multiply, thus causing infections. Second, immunity is low. Normal bacteria can enter tissues and blood. Third, the flora balance is broken.

With the efforts of scientists and all human kind, most of the pathogens have been "arrested". However, the horrible scenes in the past still haunt us. Moreover, if humans continue damaging the earth without restraints and ignore public health, there will be again the danger of those horrors. There have been many painful records of infectious diseases caused by bacteria. During our long history, each time an epidemic disease spread around, thousands of lives were deprived, and many families were broken and displaced.

During 1348–1666, bubonic plague, also known as "black death", swept Europe and caused 25 million deaths. It was one of the biggest disasters in history. In 1611, the plague spread in Constantinople, causing 200,000 deaths. In 1672, the plague broke out in Naples, Italy, taking 400,000 lives; in Lyon, France, 60,000 died of the disease. In 1711, the plague spread to Austria and Germany, causing 500,000 deaths. During 1722–1855, in Yunnan Province, China, 31 counties and cities were affected by plague, resulting in 253,000 deaths. During 1826–1837, cholera broke out in Europe, causing millions of deaths. In 1831 alone, 900,000 people died. During 1840–1862, cholera became a global epidemic. It lasted for 20 years, causing millions of deaths. During 1856–1900, up to 86 counties in Yunnan Province, China, were affected by plague, causing 730,000 deaths. During 1863–1875, cholera had been around for more than 10 years. In 1866 alone, 300,000 died in Eastern Europe. The total number of casualties reached hundreds of millions. During 1884–1953, in 57 cities and counties of Fujian, China, 710,000 died of plague. During 1887–1919, in Inner Mongolia in China, 520,000 died of plague. During 1910–1913, the plague broke out in China and India. During the epidemic, death amounted to millions. In 1921, cholera broke out in India, causing 500,000 deaths. Meanwhile, plague was also spreading around the country, causing more millions of deaths. In 1924, cholera broke out again in India, causing 300,000 deaths.

The dreadful "white powder": Bacillus anthracis

In 2011, FBI identified a man named Wagner as "a highly dangerous person who can escape from dire situations, might be equipped with heavy weapons, and announces himself as a warrior against abortion". Wagner was listed one of the "ten most wanted men". There was a reward of $50,000 to catch him. Who was this dangerous man? FBI received information during Thanksgiving. Wagner had claimed that he had sent more than 280 envelopes containing anthrax powder to abortion clinics around America. Abortion clinics in at least 12 states in America received such "anthrax" threat letters. On the envelope was signed "Army of God". Although investigation revealed that the white powder in the envelopes were not anthrax spores, there was still sufficient attention to the danger of anthrax.

Especially after 9/11, the danger of *Bacillus anthracis* being used in terrorist attacks was considerable to the public's attention. People felt terrified at hearing the word "anthrax". What are anthrax spores? How come even America, where technology is so advanced, is afraid of it? This concerns a strange disease in human history — anthrax.

The name of anthrax is derived from coal. It was named after the typical black skin of patients. It is an acute infectious disease caused by *Bacillus anthracis*, which affects both humans and animals.

Ancient peoples viewed anthrax as an "act of God". Around 1500 BC in Egypt, the fifth livestock plague and the sixth ever plague known as the "plague of boils" affecting both humans and animals was probably the earliest record of anthrax. In 25 BC, Roman poet Virgil described an animal epidemic with many similarities with anthrax, and he warned that the disease could infect humans through contaminated animal skins. Huangdi Neijing from ancient China also recorded anthrax. For quite some time, diseases reported in Chinese historical archives have been mostly anthrax. Entering the 20th century, anthrax spread through natural means remained a major threat to the human population. During 1900–1978, in America, 18 patients infected with anthrax were found: most of whom worked with wool or wool processing. This was why anthrax has always been called "shearer's disease" in America. During 1978–1980, anthrax among humans broke out in Zimbabwe; 6000 were infected and more than 100 died. Skin anthrax is most common. Early symptoms include spots on the face, neck, shoulders, hands, feet, and other exposed areas on the skin, which resemble mosquito bites. The spots feel itchy. This is followed by blisters and hemorrhagic necrosis, which later develop into ulcers. Afterward, blood secretions form black dry scabs like charcoals. Under the scab, granulation tissues form anthrax carbuncles. The scab falls off and forms a scar. During the disease, patients experience symptoms such as fever, headache, swollen local lymph nodes, and splenomegaly. In some cases, the condition is serious enough to cause circulatory failure and, subsequently, death. If pathogens enter into the blood, they can cause sepsis as well as pneumonia and meningitis.

Pulmonary anthrax is caused by inhaling anthrax spores. Clinical manifestations include chills, fever, shortness of breath, difficulty in breathing, wheezing, cyanosis, blood sputum, and chest pain. Patients are often in serious conditions, accompanied by sepsis and septic shock and occasionally by meningitis. Without timely diagnosis and rescue, patients often suffer acute symptoms within 24–48 h and die of respiratory and circulatory failures.

Intestinal anthrax caused by food infection leads to clinical manifestations such as acute gastroenteritis and acute abdomen diseases. The former may cause serious vomiting, abdominal pain, and watery diarrhea, which in most cases will recover in several days; the latter is accompanied by severe sepsis, persistent vomiting, diarrhea, bloody stools, bloating, and abdominal pain. If not treated timely, it will lead to sepsis and septic shock, from which the patient will die in 3–4 days.

Bacillus anthracis is the first pathogen discovered in human history. It is often big, and it is one of the biggest pathogenic bacilli. Its two ends are flat. There are no flagella; however, there are spores. The bacillus does not require much nutrition. It grows easily under normal conditions. It can form spores even in disadvantaged environments, thus possessing high resistance and vitality.

Bacillus anthracis possesses three powerful pathogenic weapons. First weapon is the capsule wrapped around the bacteria. Capsule can resist devouring human immune cells, thus facilitating proliferation of the bacteria. The other two "powerful weapons" are toxins: one is called lethal factor, and the other is edema factor. Both toxins are like bombs. Once one factor enters a human cell, the cell will burst and die. This causes a violent reaction in the human body, thus leading to death.

Anthrax spores can survive for a long time in nature, unhampered by the sun, heat, or normal disinfectants. Even on dead bodies from long time ago, it can still be a source of infection. In 1997, a police museum in a British town was being renovated. Among the debris, a small glass tube was discovered. To people's surprise, this forgotten item was actually captured from a German spy in 1918. Museum staff sent the tube to professional institutions. After inspection, the content was identified as *Bacillus anthracis*. The bacillus was not only alive but also pathogenic. *Bacillus anthracis* was also used as a biochemical weapon during the war. During Japan's invasion of China, the notorious Unit 731 once established a production line that produced 200 kg of anthrax powder every month. The Japanese army conducted various experiments in China and used biochemical weapons such as *Bacillus anthracis*, cholera, and *Salmonella typhi*. They had killed millions of innocent Chinese people.

Disease of the 19th century: cholera

Since 1817, there have been seven pandemics of cholera around the world, causing enormous numbers of death. Its damage back then is still shocking today's terms.

The first pandemic of cholera began in 1817. It started in India and spread to Arabic regions, Africa, and the Mediterranean; during the second pandemic in 1826, cholera spread to Afghanistan and Russia, then to the entire continent of Europe; during the third pandemic, it traveled across the ocean and reached North America in 1832. In less than 20 years, cholera has become the most horrifying and notable disease of the 19th century. In the 100 years until 1923, there had been six pandemics of cholera, the damage of which was innumerable. In India alone, more than 38 million people died. During the fifth pandemic in 1883, Koch discovered *Vibrio cholerae* from the feces of an Egyptian patient for the first time. After 1961, a seventh pandemic broke out. It started in Indonesia, and then spread to other Asian countries and Europe. In 1970, it entered Africa. The land that had been a stranger to cholera for a hundred years suffered from it again. In 1991, cholera caused considerable disturbance in Latin America. In 1 year, there were more than 40,000 cases and 4,000 deaths.

China was affected by every pandemic of cholera. During the 130 years between 1820 and 1948, there had been nearly a hundred outbursts of various severities. In some annual reports, the number of infected patients reached tens or hundreds of thousands. Death rate was as high as 39%.

Cholera is a highly infectious disease caused by *Vibrio cholerae*. *Vibrio cholerae* bacteria are curvy, like an arc or a comma. On one end of the bacteria is single flagellum and pili. Under normal conditions, only humans are susceptible to *Vibrio cholerae*. It is mainly transmitted by mouth through contaminated water or food. Under certain conditions, it enters the small intestine. With the move of flagellum, the bacterium penetrates the mucus layer of the mucosal surface and attaches to the intestinal epithelial cells with the help of pili. It then multiplies rapidly on the mucosal surface, secreting cholera toxin. This causes hyperthyroidism in intestinal mucosal cell function. Patients show symptoms such as vomiting and diarrhea. The diarrhea effluent looks like water left behind after rinsing rice and contains considerable amount of *Vibrio cholerae*. Serious dehydration and salt loss may cause metabolic acidosis. Blood circulation may fail, and the patient can suffer shock or death. Those who had once been infected obtain strong immunity; they will seldom be infected with cholera again.

Vibrio cholerae can resist low temperature and alkali, but it is considerably sensitive to heat, dryness, sunlight, chemical disinfectants, and acid. Fifteen minutes in a hot and damp environment at 55°C can kill *Vibrio cholerae*, as well as 1–2 min in an environment at 100°C or 15 min in water with a chlorine density of 0.5 g per ton of water. Potassium solution (0.1%) can disinfect vegetables and fruits.

Black demon: the plague

The plague is also known as the black death. Many pandemics of the plague have left horrifying records in human history. The first pandemic of the plague was in 6 BCE. It started in the Middle East, eventually spreading to Europe through the Mediterranean. The pandemic lasted for 50 or 60 years. At its peak, tens of thousands of people died every day. The total death count reached 100 million. This pandemic caused the recession of Eastern Roman Empire. At that time, the Justinian dynasty was ruling Ethiopia, thus the pandemic was also called the "Justinian Plague".

The second pandemic occurred in the 14th century. There are various accounts of its origin. It affected the Eurasian continent as well as the Northern African coast. This time the pandemic continued for almost 300 years. In Europe alone, 25 million had died, which was one-fourth of the entire European population at that time. Italy and Britain lost almost half of their population. According to records, there were rotting bodies of cats and dogs everywhere on London streets — they were killed as being responsible for the pandemic. Without cats, however, the real carrier of the plague, mice, became more unrestrained. Up until August 1665, 2000 people died every week. After a month, the death count reached 8000. It was not until the Great Fire of London months later, which destroyed most London architectures, as well as most mice that carried the plague. This pandemic had also affected China. In 1793, Shi Daonan of Yunnan wrote in his poem "Dead Mouse Lines" that at that time, "There are mice to the East and the West; people see them as tigers; not many days of the mice's death, men died as well". This demonstrated how serious the plague was at that time in China.

The third plague pandemic started at the end of the 19th century and reached its peak around the 1930s. It affected more than 60 countries in Asia, Europe, America, and Africa, causing tens of millions of deaths. This pandemic was far greater than the previous two for the speed at which it spread and the regions it affected. But this time, the plague was more quickly and completely controlled than previous times. The reason is that by that time, the plague pathogen, *Yersinia pestis*, had been discovered. There were preliminary understandings of how the plague started and how it was transmitted. International quarantine measures were also strengthened. Battle of humanity with the plague has entered a scientific phase.

The plague is a deadly infectious disease caused by plague bacillus (namely *Yersinia pestis*). *Yersinia pestis* is a small bacillus with a capsule. Its pathogenic substances include exotoxins, endotoxins (lipopolysaccharide, LPS), and capsule.

The plague usually breaks out first among mice and other wild rodents, and it is transmitted to humans through mouse or flea bites; humans can also be infected by direct contact with or a bite from an infected animal; between humans, pathogens can be transmitted through saliva. When *Yersinia pestis* enters the skin, its capsule prevents it from being devoured by phagocytes. It breeds in local areas and then enters lymphatic vessels with the help of hyaluronic acid and soluble celluloses. This causes primary lymphadenitis (the bubonic plague). The large amount of pathogens and toxins in the lymph nodes enter the blood, thus causing systemic infection, sepsis, and severe symptoms of poisoning. The liver, the lungs, spleen, and central nervous system can all be affected. When pathogens spread to the lungs, there will be a secondary pneumonic plague. The symptoms of the plague include high fever, swollen lymph nodes, lymphatic and systemic vascular endothelial cell damage, large-scale skin bleeding, bruising, necrosis, and inflammation. The body of a victim of the plague is purplish black; therefore, the disease is called "black death".

Yersinia pestis can survive for a long time in low temperatures or inside organisms: 10–20 days in purulent sputum, weeks to months in dead bodies, and more than 1 month in flea feces. It is sensitive to light, heat, dryness, and general disinfectants. It can be killed under the following conditions: direct sunlight for 4–5 h; 55 °C heating for 15 min or 100 °C heating for 1 min; 5% phenol solution; 5% Lysol; 0.1% mercury; or 5%–10% of chloramine.

White plague: Mycobacterium tuberculosis

Tuberculosis is a chronic disease that has been affecting our health for a long time. In the bones of early human remains dug out by scientists, a hunched spine was found — a symptom of tuberculosis. On a female body from Mawangdui, a grave from Han Dynasty 2100 years ago, tuberculosis calcification on its left lung was also discovered. Mummies infected by tuberculosis were found in Egypt. The descriptions of tuberculosis can be traced back to 460 BC. In recorded history, especially during the Industrial Revolution, tuberculosis had deprived many lives. As patients' faces are often white, tuberculosis is also called the "white plague". This also makes a distinction from the black death plague that swept Europe five centuries ago.

In the 19th century, tuberculosis was widespread in Europe and North America. It affected all walks of life. Poor people became the hotbed of tuberculosis. Many lost friends and family to this long and relentless disease. Many prominent names at that time suffered from tuberculosis as well, including Shelly, Schiller, Browning, Thoreau, Chopin, Chekhov, Yu Dafu, and the Bronte sisters. Tuberculosis had even affected the minds of poets and artists. It is not uncommon to find the following description in the 19th-century novels and dramas: "face white as a sheet, body thin to the bones, and waves of tearing coughs...". At the beginning of the 20th century, when tuberculosis was most severe, 2 million people around the world died from it every year.

In 1882, German scientist Koch discovered that *Mycobacterium tuberculosis* was the cause of tuberculosis. The bacillus is long and slightly curvy, blunt, and round at both ends. It appears as a single bacillus or arranged in branches. It requires high nutrition and grows slowly. Cultured in solid medium, a visible colony can be discovered only after 2–4 weeks.

The weapons used by *Mycobacterium tuberculosis* to invade human body are mainly lipids and proteins. Lipids are composed of the following components: first, the cord-like factor. It can link *Mycobacterium tuberculosis* in a chain in liquid medium, thus damaging mitochondrial membrane of cells, affecting cell breathing, inhibiting migration of leukocyte, and causing swollen chronic granulation. Second is phospholipid. It prompts the formation of tubercles. Third component is sulfuric cerebrosides, which enables *Mycobacterium tuberculosis* to survive for a long time in phagocytes.

Mycobacterium tuberculosis can survive for 6–12 months in low temperatures (such as 3°C); it is highly resistant to dryness and relatively resistant to acid and alkali; *Mycobacterium tuberculosis* from sputum samples can resist common disinfectants such as 0.5% Lysol solution, 5% carbolic acid solution, and 0.1% peracetic acid solution for more than 1 h; 70% alcohol can kill *Mycobacterium tuberculosis* rapidly.

Mycobacterium tuberculosis can enter and infect an organism through respiratory and gastrointestinal tracts or through a cut in the skin. It then causes infections in various organs, among which the lungs are the most common. Symptoms include coughing, sputum, hemoptysis, chest pain, fever, fatigue, loss of appetite, and other local and systemic symptoms. Bodily fluids with pathogens come out of the body when the patient coughs, sneezes, or talks loudly. Healthy people will be infected when they inhale the bacteria. This chronic wasting disease will cause progressive weight loss for the patient. *Mycobacterium tuberculosis* can affect the other parts of the body, causing diseases such as skin tuberculosis, spinal tuberculosis, meningeal tuberculosis, and bone tuberculosis.

In 1945, the advent of streptomycin indicated that tuberculosis was no longer incurable. Afterward, isoniazid, rifampin, ethambutol, and other medicines were synthesized. BCG (bacille Calmette-Guérin) vaccination was successful. This was a milestone in struggle of humanity against tuberculosis. Therefore, in the 1980s America, it was believed that by the end of the century, tuberculosis would be eliminated. However, the persistent tuberculosis had launched a new round of attack to humans. It is resurging. In many countries, numbers of tuberculosis cases have been soaring. Therefore, the World Health Organization (WHO) announced that "the world is in tuberculosis emergence" and declared March 24 to be celebrated annually as "World Tuberculosis Day".

Is leprosy the curse of God?

The earliest record of leprosy was from Ancient Egypt 3000 years ago. During 1324–1258 BC, Pharaoh Leimusaisi II ruled the vast territory of Ancient Egypt. At this time, a strange disease started spreading among black people in south Egypt and Sudan. At first, patients' hands and feet were disfigured; then their faces became distorted. In the end, they died in pain. Egyptians called the disease "sate", meaning "to fester". More than 3000 years later, modern archaeologists have finally opened the mysterious pyramids. During a careful examination of a mummy, they discovered that the poor man was tortured by leprosy in his life. His skull was badly hurt. This discovery has finally confirmed legends that leprosy had existed in ancient civilizations. It is safe to say that, as an ancient disease, leprosy has tortured humans for thousands of years.

Patients with leprosy lose their hair, including body hair and eyebrows. Their ears and noses are sometimes disfigured, making the whole face terrible to look at. Limbs can also be deformed. Patients are often shunned by their society and family. During the Middle Ages in Europe, there were often such rumors: scared people transported leprosy patients with boats and drowned them in the sea; leprosy patients were burned alive; in distant and deserted valleys, quarantine areas were built especially for exiled leprosy patients, who were treated as dead people. There was also talk that every other day a funeral would be held. Once quarantined, they are restricted from going out, and they must ring a bell or rattle wood pieces while walking, so that others could escape in time.

In ancient times, leprosy was believed to be God's punishment for those who had offended him. In Hebrew, leprosy was called "zaras", meaning "unclean and untouchable soul". Thus, leprosy patients were sinners who were discriminated against.

What is leprosy exactly? Is it really that horrible? And how is it transmitted?

In 1856, Norwegian specialist on leprosy Danielson could not resist the urge to find the truth; he took the risk to plant substance scraped from a leprosy patient's skin into him and his four assistants. Fortunately, none of the five was infected. In 1873, Hansen, son-in-law of Danielson and leprosy scholar, discovered many rod-shaped structures in the skin nodules of patients with leprosy. These were later proved to be the pathogen of leprosy — *Mycobacterium leprae*. The discovery of *Mycobacterium leprae* put an end to various ridiculous and strange tales of the cause of leprosy. Studies of leprosy thus entered the scientific era.

This bacterium is similar to *Mycobacterium tuberculosis* in its shape. It is long and slightly curvy, with the tendency to grow into branches. Under 0 °C, it can survive for 3–4 weeks; under strong sunlight, it loses reproductive ability after 2–3 h; if heated at 60 °C for an hour or under an ultraviolet light for 2 h, it loses vitality; boiling for 8 min can kill the bacteria.

Mycobacterium leprae is widely distributed in the patient's body. It can be found on the skin, in the mucous membranes, surrounding nerves, lymph nodes, liver, spleen, and other cells in the reticuloendothelial system. Bone marrow, testes, adrenal glands, and the front half of the eye are also places susceptible to *Mycobacterium leprae*. A few *Mycobacterium leprae* exist in surrounding blood and striated muscle of the above parts. *Mycobacterium leprae* can also be found in milk, tears, semen, and vaginal secretions. Its infection is primarily through respiratory tract and through contact with damaged skin. If a healthy person inhales saliva of a patient or has been in intimate contact with the patient for a long time, he/she might also be infected.

When *Mycobacterium leprae* invades the body, it first conceals in macrophages, in surrounding nerves, or tissues. The incubation period is often 3–5 years or even longer. Once infected, the outcome and the type of pathology depend on the immunity of the infected body. Tuberculoid leprosy happens mostly to the skin of the face, limbs, shoulder, back, and buttocks. There will be clearly outlined and irregularly shaped rashes or pimples that are sunken in the middle and higher on the edges. In the case of lepromatous leprosy, bacteria invade the skin, mucous membrane, and other organs, leading to granulomatous disease. If not treated in time, it will lead to death. Today, leprosy is a curable disease. Treatment in the early stage can avoid disability. For most leprosy patients who are cured, there will be "pocks" left on their faces.

Leprosy can be found worldwide. Every year around 500,000 new patients are found globally. To spread knowledge on leprosy and promote its elimination, the WHO decided in 1954 that the last Sunday of every January should be the "International Day to Combat Leprosy". Its goal is to motivate all societal resources to help patients with leprosy overcome difficulties in life and at work and acquire more rights.

Typhoid Mary

In 1869, a girl called Mary was born in Ireland. She immigrated to the United States at the age of 15. She had worked as a maid and cook. In the summer of 1906, New York banker Warren took his family to Long Island for summer vacation, and Mary was hired as a cook. At the end of August, one of Warren's daughters was first infected with typhoid fever. Next, Mrs Warren, two maids, a gardener, and another daughter were infected. Among the 11 people living in their summer house, 6 were affected. Warren was deeply anxious. He consulted the typhoid specialist Soper. Soper targeted Mary. He invested in detail Mary's working experience in the previous 7 years, and discovered that Mary had changed seven workplaces during the period. At each place there had been an outburst of typhoid. In total, there were 22 cases and 1 death.

To get the correct conclusion, Mary's blood and feces samples must be obtained. But this was tricky. Mary reacted violently. As she had been living a healthy life, the idea that she transmitted typhoid to others was an insult to her. Mary refused to be sampled. In the end, local health officials motivated five police to carry her to hospital in an ambulance. Along the way, Mary was like "an angry lion in a cage".

Hospital test confirmed Soper's hypothesis. Mary was quarantined on an island called "Northern Brothers" near New York. However, Mary did not believe the doctors' conclusion. In 2 years, she filed complaints to health departments. In February 1910, the local health department reached to an agreement with Mary and released her from quarantine, on condition that she would never be a cook again.

In 1915, typhoid burst out in a New York maternity hospital: 25 were infected and 2 died. Health officials soon found Mary in the kitchen of this hospital, who, by that time, had changed her name to "Mrs Brown". It is said that Mary was sure she was not the source of infection. That is why she resumed a job as cook, as cooks were paid decently. But this time, Mary admitted defeat. She went meekly back to the island. Doctors used all possible typhoid medicines on Mary, but the typhoid germs persisted in her body stubbornly. Mary gradually acquired some knowledge on infectious disease and cooperated with the hospital. She even became a volunteer for the hospital laboratory. In 1932, Mary had hemiplegia because of a stroke and died after 6 years.

Mary's name was embedded in American medical history in the form of "Typhoid Mary". This is the first discovery that healthy people

Figure 2-24: *Salmonella typhi* and the cause and transmission of "Typhoid Mary".

can carry pathogens. These people are called "healthy carriers". They themselves do not get ill, but they can transmit pathogens to others. What Mary carried was *Salmonella typhi*, which was discovered in 1880. It is a rod-shaped bacterium with flagella. Figure 2-24 shows *Salmonella typhi* and the cause and transmission of "Typhoid Mary".

Salmonella typhi exist in patients or carriers. They come out of the body via feces, thus contaminating water and food. They can also be transmitted through hands, flies, or cockroaches. Water pollution is the predominant way of disease transmission. It often leads to epidemic diseases. When *Salmonella typhi* enter the small intestines, they invade intestinal mucosa. Some enter the ileum through lymphatic vessels and form lymph nodes. Then the bacteria enter the bloodstream through the thoracic duct and cause transient bacteremia. Through blood, *Salmonella typhi* enter the liver, spleen, and other reticuloendothelial systems and continue to reproduce. Then, they enter bloodstream again, causing serious bacteremia for a second time, meanwhile releasing strong endotoxins, thereby causing a clinical disease. After 2–3 weeks, *Salmonella typhi* enter the intestines. Some invade intestinal lymphoid tissues again. The latter, already sensitized, will suffer severe inflammation. The complications of the intestines may include swelling, necrosis, ulceration, and even intestinal perforation. During the fourth and fifth weeks, immunity increases.

Salmonella typhi are gradually removed from the body. Tissues restore and recover. However, around 3% become chronic carriers. A few patients will be infected again due to insufficient immunity.

Clinical demonstrations of typhoid include persistent fever and sera symptoms. Other signs include roseola, typhoid tongue, relative bradycardia, and swollen liver and spleen. For most patients, body temperature rises gradually in the first week of contracting the illness; in the second week, they have persistent high fever. If the patients do not receive special antibacterial treatments, their fever will only subside in the fourth week. The whole fever can last as long as a month. Apart from the fever, patients can have other symptoms of intoxication, such as apathy, severe weariness, loss of appetite, and bloating. In most serious cases, patients demonstrate neurological symptoms such as irritability, delirium, talking gibberish, or mental confusion. After 7–10 days into the illness, many patients can find light red rash on their chest, stomach, and back. These are roseola. Other common companion diseases include typhoid hepatitis and myocarditis.

Salmonella typhi have strong survival abilities in nature. Generally, they survive for 2–3 weeks in water and for 1–2 months in feces; they can not only survive for a long time in some food such as milk and meat but also reproduce; they resist low temperatures, and they can survive for several months in a freezing environment. However, they are less resistant to light, heat, dryness, and disinfectants. Under direct sunlight, they will die in several hours, or in boiled water or heated at 60 °C for 30 min. A 5-min exposure to 3% phenol or disinfected water where chlorine reaches 0.2–0.4 mg/l can also be fatal to the bacteria.

Salmonella typhi are highly contagious. In 1812, when Napoleon invaded Russia, many soldiers were infected with typhoid. This was one of the main reasons for the defeat of France. During the Crimean War in the 1850s, the number of soldiers who died of typhoid was 10 times as many as soldiers who died in battle. In 1898, British doctor Wright developed the typhoid vaccine. The vaccine played an important role during the First World War. Millions of soldiers died of poor conditions on the battlefields; however, only 100 died of typhoid fever.

Disease that is fading from memory: diphtheria

Diphtheria is an acute respiratory infection caused by *Corynebacterium diphtheria*. It is also an ancient disease often affecting the young. Mucosal membrane of the patient's pharynx and larynx congest and swell, and there is often a white pseudomembrane at the throat surface; thus, the disease is also known as "white throat". Diphtheria used to be a horrible pandemic disease. During 1735–1740, it broke out in some cities in New England. It was reported that 80% of children under 10 years of age died from the disease. In the 1920s America, there were 100,000–200,000 patients every day, among whom 13,000–15,000 died. With the rapid development of modern medicine as well as the promotion of vaccination, diphtheria as an acute epidemic is becoming history, gradually fading from our memory. In America, for example, there had been only 57 cases of diphtheria during 1980–2004.

Corynebacterium diphtheria was discovered in 1883. It is a short stick in shape; its thickness varies by bacterium. Generally, its one or both ends swell and are rodlike. It exists in throats of patients or carriers. It is mainly transmitted through saliva in air: indirect transmission include through towels, utensils, toys, books, and newspapers. Once infected, *Corynebacterium diphtheria* breeds in the patient's nose, pharynx, and larynx and produces strong toxins, causing local pseudomembrane formation or systemic intoxication symptoms. Clinical demonstrations include nausea; vomiting; chills and fever; mucosal congestion and swelling of nose, pharynx, and larynx; slight throat pain; and difficulty swallowing. A gray pseudomembrane is formed, and it does not fall off easily. Exotoxin produced by the bacteria can cause systemic poisoning symptoms. In severe cases, the patient can have myocarditis and peripheral nerve palsy, which will lead to death. Winter and spring are the two seasons when diphtheria is most common. Fifteen-year-olds have the highest rate of being infected.

Corynebacterium diphtheria is highly resistant to dryness, low temperature, and sunlight. In a dry pseudomembrane, it can survive for up to 2 months; in water or milk, it can survive for several weeks; when spread with dust and exposed to direct sunshine, it would be killed only after several hours. However, it is susceptible to heat and chemical disinfectants. It can be killed immediately when heated at 56°C for 10 min, in 0.1% mercury, 5% phenol, or 3%–5% Lysol solution.

During 1889–1894, German army sergeant Behring focused on diphtheria research. At that time, there were 50,000 children dying of diphtheria every year in Europe. He injected *Corynebacterium diphtheria* into a type of mice and discovered that the surviving mice would not be affected when the bacteria were injected again. He believed that there was a substance in animal serum that could cure other animals. During Christmas of 1891, he successfully cured a diphtheria-infected child in a Berlin hospital with goat serum. This was a significant progress toward humanity's conquest of diphtheria. In 1892, Behring started cooperation with Frankfurt Chemical and Pharmaceutical Company. In 1894, diphtheria vaccine was produced and marked. In 1901, he was awarded Nobel Prize in Physiology or Medicine.

Pyogenic bacteria

Pyogenic bacteria can infect human body and cause purulent inflammation. They are pathogenic to the human body, often causing purulent infections of the skin, subcutaneous soft tissues, and deep tissues, as well as inflammation of internal organs. They can also lead to sepsis.

Pyogenic bacteria include two major types: pyogenic coccus and pyogenic bacillus. Pyogenic coccus includes *Staphylococcus aureus*, *Streptococcus pneumoniae*, *Neisseria meningitidis*, and *Neisseria gonorrhoeae*; pyogenic bacillus includes *Escherichia coli*, *Proteus*, and *Pseudomonas*.

Staphylococcus aureus

Staphylococcus is arranged in grapelike clusters. It has at least 20 species; while a few can cause diseases, most are nonpathogenic. *Staphylococcus aureus* is one major pathogen of this type. It causes skin, organic, or systemic purulent inflammation.

Staphylococcus aureus is everywhere in nature: air, water, dust, and human and animal feces. It is also found in the nose, throat, hair, and skin of human and animals. Therefore, it has many opportunities to contaminate food.

The virulence of *Staphylococcus aureus* is determined by the toxins and invasive enzymes it produces. It can invade an organism in various ways and cause systemic infections such as pneumonia, pseudomembranous colitis, pericarditis, and even sepsis. Adults have sufficient resistance to staphylococcal infection. But specific immunity is not as strong, thus repeated infection is possible.

Staphylococcus epidermidis

When the skin is damaged, these normal bacteria can take the opportunity to invade and cause purulent infection. In addition to *Staphylococcus epidermidis*, there are also other bacteria such as *Propionibacterium acnes* in hair follicles and sebaceous glands. When these bacteria reproduce in large numbers, there will be inflammation. The infected skin swells and is purulent, forming acne.

β-Hemolytic streptococcus

Streptococcus is spherical or oval in shape, often arranged in chains. *α-Streptococcus* is also known as *Streptococcus viridans*. It often resides in human oropharynx and respiratory and intestinal tracts. It is not strongly pathogenic. *β-Hemolytic streptococcus* or *Streptococcus pyogenes* can produce strong hemolytic toxins and is highly pathogenic. It can cause various diseases. *γ-Streptococcus* is harmless to humans.

Hemolytic streptococcus is widely distributed in nature. It exists in water, air, dust, and feces, as well as in the nose, mouth, and throat of healthy humans and animals. It can be transmitted through direct contact and airborne saliva, or infections can occur through skin or wounds in mucous membrane. Contaminated food can also infect humans. It can cause purulent inflammation of skin and subcutaneous tissues, respiratory infection, and outbreaks of pharyngitis. It can also cause sepsis, bacterial endocarditis, scarlet fever, rheumatic fever, and glomerulonephritis among the newborns.

This is not a highly resistant bacterium. Thirty minutes in an environment below 60 °C can kill it. It is sensitive to common disinfectants, as well as to penicillin, erythromycin, chloramphenicol, tetracycline, and sulfonamides. In dry dust, it can survive for several months.

Streptococcus pneumoniae

There are many causes to pneumonia, and *Streptococcus pneumoniae* are the major pathogen. They have capsule and are often arranged in pairs (*Diplococcus pneumoniae*) or in chains (*Streptococcus pneumoniae*).

Streptococcus pneumoniae usually dwell in respiratory tract as normal flora. Most bacteria are nonpathogenic or slightly pathogenic. Some are pathogenic, but those cause disease only when immunity is low. *Streptococcus pneumoniae* can cause lobar pneumonia and bronchopneumonia as well as pleurisy, empyema, otitis media, mastoiditis, sinusitis, meningitis, and sepsis, thus are very dangerous.

Streptococcus mutans

Streptococcus mutans are closely related to the formation of dental caries. They possess special affinity to dental hard tissues. They can decompose sugar into food and produce insoluble glucose with high viscosity, thus causing other floras to flourish on unclean teeth. These substances stick to teeth surfaces along with sugar and other substances, accumulating into hard plaques with a yellow-brownish color. *Lactobacillus* is also related to caries. Lactobacilli can ferment polysaccharides and produce large amounts of acid. This causes a drop in pH to around 4.5, which can corrode dental tissues. Enamel and dentin are demineralized, and shiny white teeth turn into caries.

Neisseria meningitidis

Meningococcal meningitis is a purulent meningitis caused by *Neisseria meningitidis*. Meningococcal are kidney- or bean-like in shape with spherical ends. The bacteria only exist in human bodies, with its source of infection being patients and carriers. People can be infected through saliva or contact with contaminated objects. Pathogens enter blood circulation via nasopharynx, causing sepsis. These finally accumulate in meninges and spinal cord membrane, leading to purulent cerebrospinal membrane lesions. Patients show symptoms of fever, headache, vomiting, neck stiffness, and petechial skin.

These bacteria have only weak resistance to dryness, humid heat, and cold. They die in 3 h at room temperature. They are also susceptible to common disinfectants, as well as to sulfonamides, penicillin, streptomycin, and chlortetracycline. Humans have high resistance to meningococcus. Children are more easily infected, but only 2%–3% of the infected have meningitis. Most experience nasopharyngitis or become a carrier.

Neisseria gonorrhoeae

Gonorrhea is a sexually transmitted disease, which frequently affects young people. There are long historical records of this disease. The Bible, for example, wrote about it. Its symptoms were also recorded in ancient medical books in China, such as *Huangdi Neijing — Suwen*, *Prescriptions of the Golden Chamber* by Zhang Zhongjing, and *General Treatise on Causes and Manifestations of All Diseases* by Chao Yuanfang. Gonorrhea is caused by the *Neisseria gonorrhoeae*. It leads to purulent infection of the genitourinary system. It can also affect other parts of the body such as eyes, throat, rectum, and pelvic. It can also cause systemic infection through blood circulation. *Neisseria gonorrhoeae* are round, oval, or kidney shaped. They are often arranged in pairs. They have pili. They were first discovered in 1879 by Neisser and named after him.

Neisseria gonorrhoeae have weak resistance. They die easily in dryness and prefer damp environments between 35 and 36 °C with plenty of carbon dioxide. They only survive for 1–2 h in dry environments; 18–24 h in slightly wet clothes, towels, or sheets; and only 5 min in environments at 50 °C. They can be easily killed by common fungicides.

Pseudomonas aeruginosa

Pseudomonas aeruginosa is widely found in water, air, and soil. It can also be found in intestinal tracts, respiratory tracts, and on the skin of normal people. It is a common opportunistic pathogen. It can be transmitted in many ways, but mainly through contaminated medical equipment and appliances, or through iatrogenic infection by medical staff who are carriers. Therefore, burn units, surgical instruments, and other equipment should be strictly sterilized. *Pseudomonas aeruginosa* can infect almost every tissue and parts of human body. It often starts from infections in surgical incisions or burned tissues. Symptoms include local purulent inflammation. It can also cause otitis media, keratitis, urethritis, gastroenteritis, endocarditis, empyema, as well as sepsis. It has strong resistance and is resistant to many chemical disinfectants and antibiotics. Under 56 °C, it would take an hour to kill the bacterium.

Gastrointestinal pathogens

Helicobacter pylori

In 1982, Australian scholars Barry J. Marshall and J. Robin Warren first isolated *Helicobacter pylori* from antral mucus layer and epithelial cells of patients with chronic gastritis. They tried the bacteria on themselves and proved that ulcers were indeed caused by *Helicobacter pylori*. This result shook the wrong perceptions held by then medical professionals toward ulcers and brought about a revolution in the treatment. The discovery of *Helicobacter pylori* was the biggest contribution to digestive medicine at the end of the 20th century. To award their discovery of *Helicobacter pylori* and its pathogenesis of gastritis and peptic ulcer, Marshall and Warren were awarded the Nobel Prize of Physiology or Medicine in 2005 (Fig. 2-25). By far, 80%–90% of peptic ulcers are known to be caused by *Helicobacter pylori*. Therefore, the Nobel Committee stated in its announcement, "thanks to the pioneering discovery of Marshall and Warren, ulcers is no longer a chronic disease. It can be cured with short-term intake of antibiotic and antacids".

Figure 2-25: Nobel Prize winners Barry Marshall (left) and Robin Warren (right).

Helicobacter pylori are spiral-shaped bacteria, which are very picky about their environment. By far, they are the only microorganism that can survive in the human stomach. They are long, slimy, and curvy, often in the shape of spiral, letter S, or a seagull. They are arranged like fish groups in the gastric mucous layers. Each bacterium has 4–6 flagella at one end. When cultivated *in vitro*, they reveal a rodlike shape.

Helicobacter pylori are highly infectious. They can enter stomach with food and drinking water, and then secrete adhesive substances to stick firmly to gastric epithelial cells. In this manner, they can avoid being emptied out by the stomach along with food. Meanwhile, propelled by flagella on one side, they can penetrate the mucous layer. They secrete superoxide dismutase (SOD) and catalase to protect the bacteria from leukocyte. They also secrete urease hydrolysis, which decomposes urea and produces ammonia, forming an "ammonia cloud" around the bacteria, thus protecting it against stomach acid. After settling in the stomach, the bacteria grow and reproduce, and start to corrode the gastric mucosa. This forms inflammatory lesions and finally develops into stomach diseases such as ulcers.

Helicobacter pylori attract considerable attention because they are closely related to ulcers, duodenal ulcers, and gastritis. These diseases, if not treated timely, can develop into stomach cancer. Epidemiological data show that stomach cancer often appear in groups with high rates of *Helicobacter pylori* infections; colorectal cancer, esophageal cancer, lung cancer, and other tumors do not reveal explicit correlations with *Helicobacter pylori* infection rates. This strongly suggests *Helicobacter pylori* are pathogens of stomach cancer. The WHO listed *Helicobacter pylori* as a first class carcinogen of stomach cancer.

Research show that in cancerous tissues, 69%–95% of *Helicobacter pylori* are positive; people infected by *Helicobacter pylori* face risk of stomach cancer at a rate of 2.3%–6.4%. *Helicobacter pylori* are also related to gastric-related lymphoma. Among people infected by *Helicobacter pylori*, risk of such lymphoma is 3.6 times larger than noninfected people. Therefore, once a *Helicobacter pylori* infection is cured, the risk of lymphoma can be reduced, or its development is brought under control. In addition, close relatives of *Helicobacter pylori*, *Helicobacter felis* and *Helicobacter nemestrinae* can both cause similar symptoms in mice.

Shigella dysenteriae

Shigella dysenteriae does not appear to be much different from other bacteria. It is rod shaped, with no spores, no capsule, and no flagella, but it has pili. It is named after Japanese bacteriologist Kiyoshi Shiga (1898), who first discovered it. It is the pathogen responsible for acute intestinal infectious diseases (dysentery in short), whose main symptom is diarrhea.

Bacillary dysentery is the most common intestinal infectious disease, often appearing during summer and fall. The main sources of infection include patients and carriers, and it is transmitted through food and drinking water contaminated with *Shigella*. People are susceptible to *Shigella*. About 10–200 bacteria can infect 10%–50% volunteer experiment subjects. Bacillary dysentery is widely distributed, spreads fast, and is highly infectious, thus posing great danger to human health. Especially in flooding areas, once water is contaminated, *Shigella* can easily cause and spread dysentery.

When *Shigella* enters the digestive system, it sticks to epithelial cells of terminal ileum and colon mucosal with the help of pili. Then, it enters the epithelial cells and grows, spreading to neighboring cells in the meantime. It releases endotoxins, which cause local mucosal inflammation, necrosis, and ulcers in the digestive tract. When the endotoxins enter the bloodstream, they lead to fever and increase in leukocyte. A few patients may experience toxic shock or disseminated intravascular coagulation. In addition, *Shigella* dysentery also produces enterotoxins, causing increased secretion of intestinal mucous. Common and typical symptoms include chills, fever, general malaise, abdominal pain, diarrhea, abdominal tenderness (especially in the lower left quadrant), and sticky stool. After 1–2 days into the illness, feces contain blood; excretion frequency also increases to more than 10 times a day. The symptoms last for 1–2 weeks. Later, it either heals by itself or turns into chronic symptoms.

Salmonella

Salmonella resides mainly in animal intestines. They are of many types, and they are pathogenic to both humans and animals. A *Salmonella* infection is a common foodborne illness both in China and around the world. Within the *Salmonella* family, *Salmonella typhi* and paratyphoid α, β, and γ can all lead to *Salmonella*-based typhoid and paratyphoid. It is a B infectious disease listed in the Law of China on the Prevention and Treatment of Infectious Diseases. All other salmonella apart from those mentioned earlier are called nontyphoid salmonella. Diarrhea caused by nontyphoid salmonella belongs to another type of infectious diarrhea and is listed as C infectious disease in China. *Salmonella typhimurium* and *Salmonella enteritidis* are the two most common nontyphoid *Salmonella* infection serotypes.

Nontyphoid *Salmonella* is a global epidemic. The infection rate is highest among infants and children. About 60–80% of the cases are sporadic, but it can also burst out on a large scale. In Britain, the nontyphoid *Salmonella* is listed as the top cause of foodborne diseases; in the United States, it is listed the second, following *Staphylococcus aureus*. Nontyphoid *Salmonella* infection has clinical demonstrations such as diarrhea, fever, stomach ache, and vomiting. The symptoms often last for 4–6 days. Most people can heal without using antibiotics. However, some patients, for example, children, pregnant women, elderly people, and people with low immunity, may experience more serious symptoms.

Nontyphoid *Salmonella* can contaminate food through feces of animals (e.g., cattle, horses, and mice). As nontyphoid *Salmonella* widely exist in the intestines of these animals, their feces carry the bacteria and contaminate water and food. *Salmonella* cross-contamination also exists during food production processes, for example, on knifes, cutting boards, shopping baskets, or through the chef's hands. These can even cause food poisoning.

Salmonella enters the digestive system with food. It reproduces in the small intestine and in the colon; it adheres to mucous epithelial cells and invades submucosal tissues, causing intestinal mucosal inflammation. People get sick suddenly in 12–24 h. Its symptoms include headaches, dizziness, vomiting, abdominal pain, and chills. Stool is yellow or yellowish green, with mucus and blood. Daily excretion ranges from 3–4 times to more than 10 times. Healthy adults can recover in 2–5 days; weak and elderly people may suffer for several weeks before recovering.

The degree of *Salmonella* infection is determined by the number and type of *Salmonella* taken in as well as the health status of the body. Clinical demonstrations will appear only when the number of *Salmonella* is greater than 100,000; when the number is too low, then the patients will be asymptomatic carriers. However, for children, the elderly, and the physically weak, smaller amount of bacteria can also cause clinical symptoms.

Clostridium perfringens

Clostridium perfringens was found to be the pathogen in 1892 that caused *Clostridium welchii* poisoning. *Clostridium perfringens* is an anaerobic bacterium, is one of the most common pathogens for foodborne gastroenteritis, and can cause typical food poisoning.

Clostridium perfringens is blunt on both ends and has a rodlike shape. Oval spores are located in the middle or close to the end of the bacterium and are larger than the body itself. Some bacteria have spores and capsule.

If one has eaten food contaminated by *Clostridium perfringens*, he/she will have severe abdominal cramps and diarrhea within 8–22 h. The duration of the disease is often less than 24 h. However, for some individuals, insignificant symptoms may last for 1–2 weeks. Reports also reveal that some patients die because of dehydration and other mixed infections.

Clostridium perfringens was once used by the Japanese military as a biological weapon during the Second World War. It was a dangerous bacterium that caused catastrophes.

Toxigenic obligate anaerobic bacillus

Obligate anaerobic bacteria do not like oxygen. They cannot grow when oxygen is present. Even when exposed to oxygen during sample acquisition and transportation, these bacteria die quickly. *Clostridium tetani* and *Clostridium botulinum* are both obligate anaerobic bacteria. These bacteria have spores that are often wider than the body, thus the bacteria swell to a shuttle shape. Both bacteria can produce strong exotoxins that cause diseases to humans and animals.

Clostridium tetani

Clostridium tetani was discovered in 1884. It is a slender and long rod-shaped bacterium. Most of the time, it has flagella and can move around. It produces spores after 1 day of cultivation. These spores swell into balls and move to the top of the body, thus the bacterium appears like a drum hammer. *Clostridium tetani* is widely distributed in nature and is the pathogen to tetanus. As its spores are highly resistant, *Clostridium tetani* can survive for decades in soil. When *Clostridium tetani* and its spores invade human body, spores germinate and the bacterium body produces toxins that cause disease. However, *Clostridium tetani* is an anaerobic bacterium and cannot grow in wounds. Anaerobic environment is an important condition for *Clostridium tetani* infection. Narrow and deep wounds (e.g., stab wounds) mixed with dirt or foreign objects, local tissue necrosis, or its accompanying purulent infections are situations that easily provide anaerobic environments. The pathogenic factor of *Clostridium tetani* is tetanospasmin, which is a neurotoxin. It has high toxicity, second only to botulinum toxin. *Clostridium tetani* grows only in contaminated local tissues. Generally, it does not enter the bloodstream. When tetanospasmin is produced in local tissues, it is transmitted by neurons or absorbed by lymphs, and then it reaches the central nervous system through the bloodstream. It damages the neurotransmission between neurons that normally suppresses impulses, leading to abnormal excitability. Extensor and flexor contract at the same time, causing spasm in striated muscles of the entire body. Early symptoms include fever, headache, malaise, muscle pain, and other prodromal symptoms. Meanwhile, there is also muscle twitching, difficulty in opening the mouth, spasm in masticatory muscle, trismus, and contorted facial expressions. Later, tonic contraction occurs on neck, trunk, and limb muscles. The body becomes reversely contorted like an arrow. Facial color is unnatural, and there is difficulty in breathing. In the end, patients may die of suffocation.

If we have a cut, especially a deep one, what should we do? We must go to the hospital for a vaccination. The injected vaccine is tetanus vaccine, which protects us from *Clostridium tetani* infection.

Campylobacter jejuni

Campylobacter jejuni is the most common pathogen to sporadic bacterial enteritis. Its cell is long and slim, curvy, spiral, letter S, or seagull-like in shape. There is a single flagellum on one or both ends. It only survives for 3 h in dry environments and for 2–24 weeks at room temperature.

Campylobacter bacterium exists in many animals. It causes diarrhea, gastroenteritis, and parenteral infection to humans and animals. It is present around the world, with increasing cases of infection in recent years. Currently, it has become the most common pathogen causing bacterial diarrhea. It is taken in with contaminated food, milk, or water, or is transmitted through direct contact with animals. As *Campylobacter* is highly sensitive to stomach acid, it causes disease only when more than 100,000 of it are taken in. The bacterium reproduces on the inside of the small intestine, invades the intestinal epithelium, and causes inflammation. Clinical demonstrations of infection include abdominal cramps, diarrhea, bloody or jamlike stools in large quantities, headache, malaise, and fever. The duration of the illness is 5–8 days.

Clostridium botulinum

Clostridium botulinum was discovered in 1896. It has parallel sides, blunt and rounded ends, and rodlike or slightly curvy shape. It has 4–8 flagella. The oval spore near the end provides the bacterium a shape like that of tennis racket. Sometimes the bacterium is arranged in long strides or chains, sometimes in a boat-like shape, lemon-like shape with handle, snakelike thread, or bulbous-like shape with deep dyes.

Clostridium botulinum is widely found in nature. Its toxins can survive for years in dry, sealed, and dark environments, and they are colorless, odorless, tasteless, and difficult to detect. It is sensitive only to heat. Its virulence can be destroyed quickly at 80 °C. Sausage, ham, fish and fish products, and canned food are major sources that cause botulism.

Botulinum toxin is highly toxic. It is one of the strongest nerve palsy toxins. An amount of toxin with the side of a piece of sesame can kill 2000 mice; 1 g of this toxin can kill 4 million mice; 1 µg is lethal to humans; it is terrifying.

Botulinum toxin is neurotropic. Upon entering the organisms, it acts on the brain and the muscle joints of its surrounding nerve ends, inhibiting the release of acetylcholine and resulting in muscle paralysis. Botulism is a bacterial food poisoning caused by accidental intake of food with the botulinum toxin. Botulism is not common among humans. However, once infected, it is acute. Its incubation period is short, normally 6–36 h, and 60 h at the longest. The predominant symptoms include impaired vision, general weakness, trismus, difficulty in lifting head, breathing, and swallowing, and dilated pupils.

Zoonotic pathogens

Erysipelothrix rhusiopathiae

Erysipelothrix rhusiopathiae is a slim and slightly curvy bacillus that can cause erysipelas. Erysipelas is a zoonotic disease affecting many mammals and birds. Swine erysipelas concerns humans the most. Pigs are the main host of *Erysipelothrix rhusiopathiae*. At least 30% of pigs carry *Erysipelothrix rhusiopathiae* in their tonsils. Therefore, those in the pork industry are likely to be infected. Swine erysipelas is widespread around the world, and China is one of the most affected countries. Once infected, symptoms include a skin rash, high fever, and sepsis. Sometimes, infection can lead to death. *Erysipelothrix rhusiopathiae* polyarthritis is also called stiff lamb disease. It is a chronic infection to lamb limbs. Its symptoms include long-term limping and slow growth. This disease was most common during the 1930s and 1940s; afterward, it became less common. This might be due to the increasing attention to umbilical cord disinfection of newborn lambs and castration and tail docking using rubber bands.

Listeria

Listeria is a zoonotic pathogen. It exists worldwide, and there is a trend of increasing human infections. It was isolated in 1926 from rabbits. British surgeons maintained detailed records of clinical demonstrations of the infection. In 1940, this bacterium was renamed *Listeria monocytogenes*.

Listeria monocytogenes belongs to the *Lactobacillus* family. Infections from *Listeria monocytogenes* include meningitis, septic granuloma, enlargement of lymph nodes, etc. The pathogen enters the human body through inhalation, food, or direct contact with contaminated objects. It can also be transmitted from person to person. Pathogens in infected pregnant women can be transmitted to the fetus or the newborn baby through the placenta or birth canal. This is a disease with high mortality. Sometimes, even when patients receive antibiotic treatment, death rate still remains 30%.

Brucella

Brucella is a short Gram-negative bacillus. It can cause brucellosis, a zoonotic infectious disease. There are many types of brucellosis. The sources of human infection are mainly sheep, cattle, and pigs. Most people are susceptible to the disease. Repeated infection is possible. It can also turn into a chronic disease. Once affected, there are long-term symptoms of fever, sweating, pain in joints, swelling in the liver and spleen, and other symptoms.

Brucellosis has high resistance and can survive for 4 months in organs and secretions of sick animals, and around 2 months in food. It is also resistant to low temperatures. However, it is less resistant to heat and disinfectants. It is sensitive to streptomycin, tetracycline, and chloramphenicol. Infection can be treated with streptomycin and tetracycline or with cotrimoxazole and streptomycin.

Francisella tularensis

Francisella tularensis can cause the zoonotic disease, tularemia. In 1907, Ancil Martin diagnosed the first patient with the disease in America; in 1911, George McCoy and Charles Chapin first isolated the pathogens from ground squirrels in Tulare County, California. In 1921, Edward Francis named the disease tularemia.

Francisella tularensis can infect more than 250 types of domestic and wild mammals, birds, reptiles, fish, as well as humans. It can be transmitted through saliva, direct contact, food, or insects. Humans can be infected through food or through contact with infected animals. Inhaled bacteria can cause lung infections; direct contact with or ingestion of infected carcasses of wild animals can lead to ulcerative small gland, eye glands, mouth or throat lymphadenitis, or typhoid-like infections. Almost all patients can be cured. The death rate of untreated patients is 6%. Death is usually due to infection, pneumonia, infection of the meninges (meningitis), and abdominal infection (peritonitis). Proper treatment can prevent a relapse of the disease. A person, after being cured from the infection, will be immune to the disease.

There are four types of *Francisella tularensis*. The most common one (ulcer gland) causes ulcers on hands and fingers. Lymph nodes on the same side of the infection will be swollen. The second type (meibomian gland) causes red and swollen eyes, as well as swollen local lymph nodes. This type is probably infected by finger contact with the eyes. The third type (glandular) causes swollen lymph nodes but not ulcers. Generally, its cause is the intake of bacteria with food. The fourth type (typhoid) causes fever, stomach ache, and consuming symptoms. If the bacterium enters the lung, it can cause pneumonia.

Streptococcus suis

Streptococci are conditional pathogens. There are many types that exist in nature, especially among pigs. Around 30%–75% pigs are carriers of the bacteria, but not all of them are sick. High temperature, high humidity, change in temperature, and unsatisfactory sanitary conditions can all trigger *Streptococcus suis*, which is an acute, infectious zoonotic disease. It can lead to septicemia of pigs, pneumonia, meningitis, endocarditis, and arthritis. It can also be transmitted to humans through wounds and digestive systems, leading to death.

Recently, there have been reports of people dying of *Streptococcus suis* in many countries in Europe, the Americas, and Asia. In China,

Streptococcus suis is listed as a B-type animal disease. Since the establishment of the People's Republic of China in 1949, the disease had broken out in Guangdong, Jiangsu, and many other provinces. Among them, there was an epidemic outbreak in Nantong, Jiangsu, causing many deaths. In 2006, *Streptococcus suis* broke out in Ziyang and other regions of Sichuan province, causing many pigs to die. The disease also affected humans. Among the 200 cases, 37 died.

Legionella hidden in water pipes

Legionella is an infectious pathogen. It can cause acute respiratory infections, namely Legionnaires' disease. Legionnaires' disease is a new type of infectious disease emerging along with the development of human societies and the growing economy.

Legionnaires' disease first broke out during a veteran's assembly in Philadelphia in 1976. During the assembly, there was a pneumonia epidemic. There were 221 people sick, and 34 died. This attracted much attention. Since the cause of the disease was unknown, and all patients were veterans, the disease was also known as "veteran disease". The following year, a new pathogen was isolated from the lungs of dead patients. In 1978, this pathogen was internationally named *Legionella*. Later, *Legionella* was found in Europe, Australia, and other countries and regions. Each outburst had drawn considerable attention. Therefore, the WHO included the disease in the scope of disease transmission and reportage. In 1982, *Legionella* was first confirmed to be present in Nanjing, China. Afterward, there had been many reports of the spread and outbreak of Legionnaires' disease. *Legionella* is an aerobic Gram-negative bacillus. It does not form spores and has no capsule. It prefers heat and is sensitive to cold environments. It is very difficult to grow the bacterium artificially, which is why it was not discovered for a long time.

Legionella hides in all types of waters, especially in cooling water of air conditioners and artificial pipe water. In normal tap water, it can survive for more than 400 days. In natural water of 31–36 °C, it can survive for a long time. It can even be found in 60 °C environments or even in ponds near volcanoes. With the improving living standards, hot water pipes, spray sinks, showers, humidifiers, and other equipment are becoming more common. All these can add to the risk of *Legionella*. *Legionella* is the most lethal killer assembled in air conditioners. Therefore, Legionnaires' disease is acknowledged globally as the "disease of modern, civilized life".

So, how does *Legionella* in water come to inhabit human lungs? *Legionella* is mainly infected through inhalation. Aerosol is the main carrier of *Legionella*. Water supply systems can form aerosols from artificial fountains, taps, whirlpool baths, and bubble baths. Once aerosols carrying *Legionella* are blown into the air by air conditioners, they will be inhaled while breathing. *Legionella* will have the opportunity to invade alveolar tissues and macrophages, causing inflammation and collectively Legionnaires' disease.

Legionnaires' disease is slow to emerge. In general, it has 2–10 days incubation period before an abrupt onset. Once this happens, patients will show symptoms of fever, headache, chills, cough, chest tightness, fatigue, and other symptoms similar to those of the upper respiratory tract infections. In severe cases, patients experience neuropsychiatric symptoms such as retardation and delirium. They can also have respiratory failure and shock. If an epidemic of Legionnaires' disease occurs, it will be highly dangerous, with a death rate of 5%–30%.

New enemies: Escherichia coli O157, Escherichia coli O104:H4, "superbug", …

Escherichia coli variant O157

Escherichia coli probably affects humans the most. This is a short and small bacillus, which is blunt and round on both ends. It appears like a finger. Mostly, it has pili, which is thinner, shorter, straighter, and more numerous than flagella. Some cells have a capsule. *Escherichia coli* is a normal dwelling bacterium in human and animal intestines. It was discovered by Escherich, thus it is also known as intestinal *Escherichia* bacteria. Generally, *Escherichia coli* exists in the intestines of humans and all types of animals. It is a normal flora of the human body. It can inhibit the growth of harmful bacteria and synthesize vitamins. It is excreted with stool and is thus widespread in nature. Therefore, if *Escherichia coli* is detected, then there is direct or indirect contamination by stool. Thus, the unit number of *Escherichia coli* is used as hygiene bacteriological indications for drinking water, milk, and food.

Most *Escherichia coli* are harmless to human bodies. But *Escherichia coli* can cause diarrhea. It is thus known as pathogenic *Escherichia coli*. O157 is a variant of *Escherichia coli*. O is the first letter of the name of this bacterium in German. There are by far 173 types of *Escherichia coli* based on their different reactions to antigens and antibodies. O157 is thus named because it was discovered in America in 1982 as the 152th type. People used to believe that O157 inhabiting human intestines were harmless. But there had been many reports of disease caused by *Escherichia coli* O157. It was not until 1982 that scientists realized that O157 can release toxins, thus concluding that it is related to intestinal diseases. Ten *Escherichia coli* O157 is enough to cause disease. In 1982, cases of hemorrhagic colitis were reported in Oregon and Michigan in the United States. American doctors first isolated O157 from the stool of a patient. After investigation, they discovered that all these patients fell ill after eating at the same fast-food chain. The O157 epidemic in Osaka, Japan, in July 1996 was probably the most severe in history. Back then, in 62 public primary schools, whose food was provided by the same fast-food company, 6351 were infected. The epidemic soon spread to other 40 counties and prefectures in Japan, infecting almost 10,000 in total. In September 2006, spinach in America were infected by O157, affecting half of America. Later, investigation showed that feces from a cattle manure contaminated water that was used for spinach cultivation, thus causing the "poisonous spinach" in America.

The incubation period of *Escherichia coli* O157 is usually 3 days. Patients often have symptoms such as stomach aches, fever, dehydration, or bleeding diarrhea. In serious cases, patients experience hemolytic uremic syndrome and encephalitis, which can be life-threatening. By far, Japan, Canada, Switzerland, and other countries have listed O157 as a most-reported infectious disease. *Escherichia coli* O157 often attaches to the internal organs of livestock. It can resist freezing and can breed at 20 °C. Under human body temperature, its breeding capacity increases four times. However, it is not resistant to high temperature and dies at 75 °C. Heating food is an effective way to prevent O157 infection.

EHEC O104:H4

According to information released by the health department of Germany in early 2011, there was an outbreak of enterohemorrhagic *Escherichia coli* (EHEC) infection in Germany. The source of EHEC O104:H4 infection is not yet clear. An EHEC strain was detected in the suspected Spanish cucumbers but it was apparently different from the O104:H4, thus was not considered the culprit of this epidemic. Cases of hemolytic uremic syndrome were also reported in Austria, Denmark, Germany, France, New Zealand, Norway, Sweden, Switzerland, and the United Kingdom. According to a report by Associated Press from Germany, by June 1, 2011, there were 1534 cases reported, among whom 470 displayed hemolytic uremic syndrome and 17 died. Many countries issued warnings for cucumber, tomato, and lettuce consumption. In Germany, 61% of the patients were female, 88% were aged 20 and above, and around 88% had hemolytic uremic syndrome.

Superbug

The so-called "superbugs" are those bacteria that have developed resistance to clinical antibiotics. The August 2010 edition of the *Lancet Infectious Diseases*, a world-renowned journal published in Britain, stated that a new super bacterium is spreading among some countries. According to the records, a 59-year-old Indian Swede returned to India in November 2007 and received surgery in a hospital in New Delhi on December 2007. On January 8, 2008, he returned to Sweden. During his stay at the New Delhi hospital, he used antibiotics such as amoxicillin, amikacin, gatifloxacin, and metronidazole. On January 9, 2008, *Klebsiella pneumoniae* was isolated from his urine. Later, it was discovered that this bacterium is resistant to most antibiotics; it carries metal β-lactamase, and it was named New Delhi-Metallo-1 (NDM-1). The newly discovered NDM-1 bacterium is almost resistant to all antibiotics, thus it is also known as NDM-1 superbug.

The role of the metal β-lactamase is to destroy β-lactamase antibiotics often used clinically, including penicillin. These antibiotics have a ring structure that can prevent bacteria from replicating, thus eliminating bacteria. Metal β-lactamase destroys this ring structure of antibiotics. Bacteria with metal β-lactamase are thus resistant to antibiotics. What is concerning is that the genetic information of NDM-1 is able to transmit among bacteria. NDM-1 gene exists in a movable plasmid with other resistance genes and is easily transmitted to other bacteria. Today, the NDM-1 superbug has been found in France, Belgium, America, Canada, Australia, Japan, and other countries. The speed at which it spreads is alarming. With globalization, international tourism and medical tourism both contribute to accelerating its spread. The NDM-1 gene is also likely to reach places around the world in a short time. The current term of "superbug" generally refers to "ESKAPE". These six letters represent six well-known anti-medicine bacteria: vancomycin-resistant enterococci, methicillin-resistant *Staphylococcus aureus*, *Klebsiella pneumoniae*, *Acinetobacter baumannii*, *Pseudomonas aeruginosa*, and Enterobacteriaceae. Therefore, researchers warn that the spread of these bacteria might indicate the end of the era of antibiotics.

New culprit of obesity: Enterobacter

Professor Jeffrey Gordon of Washington University and Professor Miriam Karni from Katholieke Universiteit Leuven, respectively, conducted experiments in 2004 and 2007, making progress in the studies of correlation between intestinal flora, fat metabolism, and insulin resistance. However, precisely, what type of bacteria can control the genetic expressions of fat metabolism and release endotoxins that cause obesity and inflammation remains an unsolved question.

Chinese researcher Zhao Liping and his team made new discoveries in their research in the past years. They injected an *Enterobacter cloacae* isolated from the intestines of obese mice into noncarrier mice. The result was serious obesity and insulin resistance in the noncarrier mice. Decrease of such bacteria can cause decrease in weight. This provides direct empirical evidence to "chronic intestinal theory", which argues that intestinal flora is related to occurrence and development of obesity and diabetes. Further clinical studies show that in conditions that can produce endotoxins led to an overgrowth of *Enterobacter cloacae* in the intestines of an obese patient weighing 175 kg, making up one-third of the total amount of bacteria in the body. With the intervention of a carefully designed diet, the number of this bacterium reduced to an almost undetectable level. The patient's weight dropped to 51.4 kg in half a year. Symptoms of high blood sugar, high blood pressure, and high cholesterol also reduced. This is the first time causality between intestinal flora and obesity is proven.

Zhao Liping's research found that *Enterobacter cloacae* could produce endotoxins. These toxins can cause severe obesity in mice that did not gain weight with high-fat diet. They also cause inflammation and insulin resistance, shut down genes required in fat metabolism, and activate fat synthesis genes. *Enterobacter cloacae* are the bacteria that can cause obesity that the world has been looking for.

During the experiment, Zhao Liping and his team followed the Koch Law, which states that certain bacterium is the cause of an infectious disease. They first decided a certain bacterium is related to obesity; then they isolated the bacterium and replicated the disease in animal models. This demonstrates that such bacterium is the cause rather than result of obesity. This work opens up new possibilities and provides new technologies for isolating and identifying more bacteria that affect obesity and diabetes. Furthermore, we can demonstrate how intestinal flora cooperates with diet and the mechanisms that cause obesity and diabetes. There is hope for new methods to prevent and treat obesity and diabetes which target intestinal flora.

Chapter Three

A Silent Battle in the Body

What happens in the invisible front when pathogens invade the human body? Now, let us explore the smokeless immunity battle of invasion and anti-invasion.

"Route" of Bacteria's Invasion into Human Body

Different pathogens invade the human body through different routes. The respiratory tract, digestive tract, skin wounds, and urogenital tract are the main routes for invasion (Fig. 3-1).

From the respiratory tract

The respiratory tract is the communication channel for the inner part and outer part of the body. It is also the "boulevard" for bacteria invasion. Mycobacterium tuberculosis, diphtheria, pertussis, and pneumococcus, and other pathogens enter the body with air through the respiratory tract, causing tuberculosis, diphtheria, whooping cough, and pneumonia. These patients are also "sources of infection." Many bacteria reside in their respiratory tracts. When the patients cough, spit, sneeze, or talk loudly, bacteria will splash everywhere with saliva.

Invasion from digestive tract

As the saying goes, "disease enters from the mouth." The digestive tract is another "boulevard" for bacteria invasion. The pathogens of gastrointestinal diseases include *Shigella, Helicobacter pylori, Salmonella*

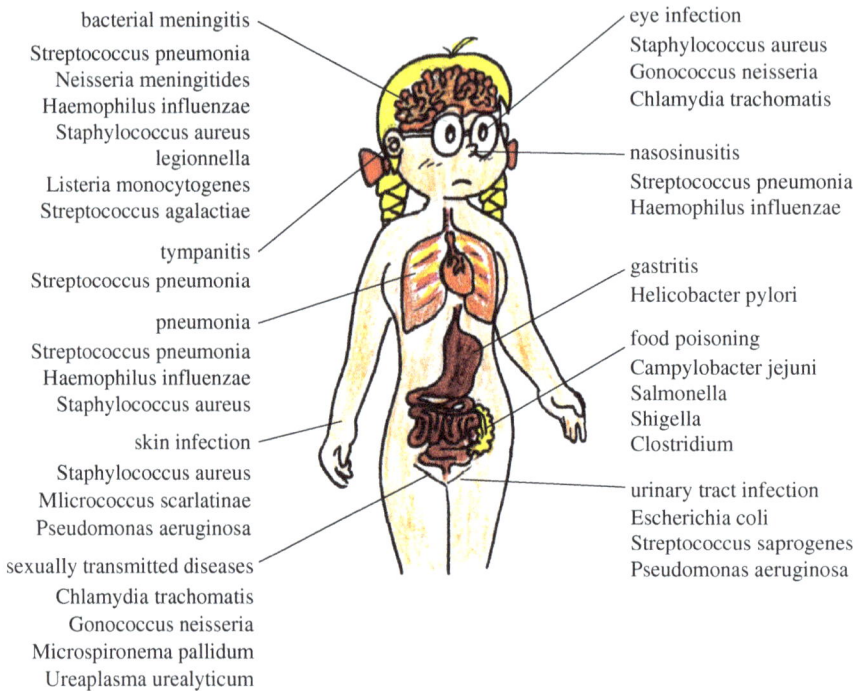

bacterial meningitis
Streptococcus pneumonia
Neisseria meningitides
Haemophilus influenzae
Staphylococcus aureus
legionnella
Listeria monocytogenes
Streptococcus agalactiae

tympanitis
Streptococcus pneumonia

pneumonia
Streptococcus pneumonia
Haemophilus influenzae
Staphylococcus aureus

skin infection
Staphylococcus aureus
Mlicrococcus scarlatinae
Pseudomonas aeruginosa

sexually transmitted diseases
Chlamydia trachomatis
Gonococcus neisseria
Microspironema pallidum
Ureaplasma urealyticum

eye infection
Staphylococcus aureus
Gonococcus neisseria
Chlamydia trachomatis

nasosinusitis
Streptococcus pneumonia
Haemophilus influenzae

gastritis
Helicobacter pylori

food poisoning
Campylobacter jejuni
Salmonella
Shigella
Clostridium

urinary tract infection
Escherichia coli
Streptococcus saprogenes
Pseudomonas aeruginosa

Figure 3-1: Route for bacteria invasion.

typhi, and *Vibrio cholerae*. All these pathogens enter the human body using this route.

Invasion from skin wounds

If the skin is damaged, the wound will provide another route for bacteria invasion. *Clostridium tetani*, for example, enters the body through broken skin. It then reproduces in the anaerobic environment in deep wounds, causing infection.

Invasion from urogenital tract

Neisseria gonorrhoeae invades through the urogenital tract, causing urethritis and cervicitis. If not treated promptly, it may also invade other parts of the urogenital tract, leading to epididymitis, prostatitis, salpingitis, and pelvic inflammation.

Invasion through insects

Sometimes, insects and animals can be the media for bacteria transmission. *Plague bacillus*, for example, is transmitted through fleas on mice. There is not only one way for bacteria invasion. Some pathogens, *Bacillus anthracis*, for example, can enter the body through various routes including the respiratory tract, digestive tract, and skin wounds. It then causes local or systemic diseases.

Bacteria's Invasion Strategy

What are the strategies pathogens adopt to break through the human immune system?

Assault on "camp" by pili the "whiskers"

After an outdoor camping trip in the fall, you will find cocklebur fruits hanging on your coat. If you look carefully, you will see the small hooks on the tip of their thorns, which allow them to stick to the clothes. Pili of pathogens have similar functions. With the help of pili, bacteria selectively adhere to certain epithelial cells and mucous membranes in the body. The pili of Neisseria, for example, allow the bacteria to adhere to the surface of the urethral mucosa, so it will not flow out with urine. *Escherichia coli* can adhere to intestinal epithelial cells with the help of its pili. Streptococcus can adhere to oral mucosa. Pathogens without pili can be easily removed with respiratory cilia movement, bowel movement, or mucus secretion. Therefore, once a bacterium loses its pili, it also loses part of its pathogenic abilities. In addition, the bacteria secrete some sticky macromolecules; these macromolecules would bind to specific surface molecules of human cells, thereby firmly sticking to the invasion site.

After pathogens adhere to mucosa or epithelial cells of respiratory, digestive, or urogenital tracts, some use these places as a "base" and start reproducing, causing infections; some penetrate cells and mucosa, reproducing inside the epithelial cells, causing superficial tissue damage; and some continue to penetrate deeper, causing diseases.

Defense against attacks from "our army" with the "bulletproof vest" capsule

Some pathogens such as pneumococcus and anthrax have sticky or gel-like capsules on the outside. With the protection of capsules, bacteria are equipped with bulletproof vests. They can resist all weapons of the immune system. This is the first step of bacteria resisting and evading the attack of "our army."

Bacteria capsules have two main ways to resist and evade. One is anti-phagocytosis: capsules are hydrophilic. They occupy space and function as barriers; they can effectively resist phagocytosis in human bodies. The second is resistance to injury: capsules can protect bacteria from being damaged by bactericidal or bacteriostatic substances (such as lysozyme) (Fig. 3-2).

Release Enzyme "missiles" to penetrate our "fortress"

Normally, body tissue has a rigid structure. It can inhibit the unlimited spread of pathogens, so it is like the Great Wall of China. However, while the virtuous rises by one, the vice rises by ten. Pathogens are also equipped with many "missiles" to break through the tissue fortress. The so-called "missiles" are the various invasive enzymes produced by the pathogens. These enzymes can hydrolyze tissues, cells, and protein, making tissues looser and more permeable, thus facilitating rapid spread of pathogens.

Hyaluronidase

Hyaluronic acid is a polysaccharide, transparent, and gel-like substance. It fills in the space between cells and collagen and covers the tissues, acting as the "glue." *Streptococcus pneumoniae, Staphylococcus*

Figure 3-2: Bacteria can resist phagocytosis and injury.

aureus, and other pathogens can secrete hyaluronidase, which increases cell gaps in the originally sturdy tissues, making it as loose as a sponge. Pathogens can move freely in the "sponge," and even spread to the entire body. Figure 3-3 demonstrates how hyaluronic acid is damaged by hyaluronidase produced by bacteria.

Collagenase

Collagen exists in skin, bones, teeth, tendons, and other parts of the human body. It is the "bonding substance" for tissues. Some Clostridium can secrete collagenase that hydrolyzes collagen in muscles and subcutaneous tissues. This loosens tissue structure, thus facilitating the spread of bacteria.

Streptokinase

To resist pathogens invasion, the human body usually forms fibrin clots where bacteria invasion most often occurs. However, some pathogens can produce streptokinase. Streptokinase can turn

Figure 3-3: Schematic for hyaluronic acid being damaged by hyaluronidase produced by bacteria.

plasminogen in fibrin clots into plasmin enzymes. The latter then dissolves the fibrin clot. After the barrier is removed, bacteria can penetrate further into the tissues.

Streptodornase

Human cells release the DNA they contain upon their death. Thus, a part of it becomes sticky and pus-like. The stickiness makes it difficult for bacteria to move freely and spread to other places, as it is difficult for a bug to move around in sticky glue. Some pathogens produce streptodornase, which can hydrolyze DNA into smaller fragments and decrease its viscosity. The pus would become less thick, making it easier for bacteria to spread.

Hemolysin

Many pathogens can produce hemolysin. Hemolysin can damage human cell membranes, thus dissolving and killing the cells.

How do Bacteria Cause Disease—Lethal Weapons

The objective of bacteria's above strategy is to destroy certain tissues of the human body and reproduce there. Such damage as well as large numbers of pathogens can cause infection. In addition, pathogens possess a lethal weapon that causes disease: toxins. Toxins produced by pathogens include exotoxin and endotoxin.

Exotoxin

Exotoxin is the toxin synthesized during the growth of many pathogens. It is not only secreted in the areas surrounding the bacteria, but it can also spread to further parts of the body from the infection spot. They then selectively act on certain tissues of the body, causing all types of diseases (Fig. 3-4).

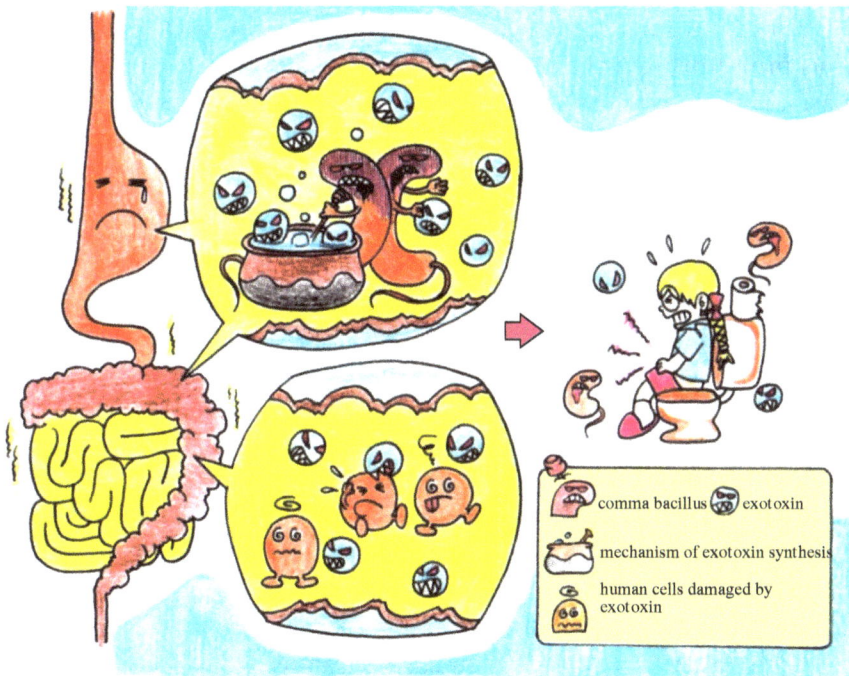

Figure 3-4: Exotoxin pathogenicity.

Exotoxins are mainly proteins produced by Gram-positive bacteria, such as *Corynebacterium diphtheriae, Clostridium tetani, Shigella,* and *Vibrio cholerae.* There are many types of exotoxins, and their pathogenicities vary. Their affinity and reaction to human cells are also different.

Tetanus toxin

Clostridium tetani is an anaerobic bacteria residing in soil. When we accidentally cut our hands or feet, and leave a rather deep wound, or when we are hurt by a rusty nail, *Clostridium tetani* will take the opportunity to enter through the wound and reproduce there, releasing a type of exotoxin—Tetanus toxin.

Tetanus toxin has high affinity to the central nervous system. Upon entering the system, this toxin will adhere to synapses and block the release of glycine. Glycine helps in muscle relaxation. Therefore, tetanus toxin can cause simultaneous strong contractions in motor neurons of extensor and flexor. Muscles suffer from spasm, rigidity, convulsions, and even paralysis. If the spasm is in the oral muscles, it will affect talking and eating. If it is in the respiratory muscles, it will cause death due to suffocation. Tetanus toxins are highly toxic. The result of its combination with neural synapses is irreversible. Normally, there are no effective treatments. Therefore, in these circumstances, one must be vaccinated in hospitals in order to prevent tetanus caused by *Clostridium tetani.*

Botulinum toxin

Exotoxin produced by *Clostridium botulinum* is called Botulinum toxin. It prevents the release of acetylcholine at nerve endings, which transmit information, thus influencing the transmission of nerve impulses, leading to flaccid paralysis in the muscles. Symptoms include ptosis, diplopia, and difficulty in swallowing. More serious cases may include respiratory muscle paralysis, which causes death due to the inability to breathe. Botulinum toxin is the most toxic substance known so far. 1 mg of pure Botulinum toxin can kill

20 million mice or 1 million guinea pigs. Its toxicity is 100,000 times greater than potassium cyanide. Mortality of Botulinum toxin poisoning is almost 100%. However, the death rate can be reduced through timely application of antibiotics and artificial respiration.

Botulinum toxins block media transmission between nerves and muscle. In recent years, people have been making use of this mechanism in facial plastic surgery to relax over-contracted muscles and paralyze overdeveloped muscles.

Diphtheria toxin

Diphtheria is transmitted through the respiratory tract. It reproduces in the epithelial cells at the mucosa of nasopharynx and releases Diphtheria toxin. Diphtheria toxin is a highly toxic cell toxin. It accumulates in places such as peripheral nerve endings and heart muscles, interrupting protein synthesis in those places. Pharynx and larynx of the patient will display symptoms such as mucus congestion, swelling, and grey pseudomembranes. In severe cases, patient may suffer from myocarditis and peripheral nerve paralysis.

Cholerae toxin

Vibrio cholerae enters the digestive system of an organism through food and drinking water. It adheres to mucosa of the small intestines, reproduces there, and secretes cholera toxin. Cholera toxin enters the epithelial cells of the small intestines, increasing the activity of adenylate cyclase in the cells, thus stimulating secretion of intestinal mucosa. Ion balance in the intestines is thus broken, causing much intestinal fluid flowing into the coelenterazine. This leads to diarrhea. In serious cases, patients suffer from vomiting, diarrhea, dehydration, metabolic acidosis, and even shock and death.

Endotoxin

Endotoxin is the lipopolysaccharide in the outer membrane of Gram-negative bacteria. As lipopolysaccharide is part of the cell, it

Figure 3-5: Endotoxin pathogenicity.

cannot be secreted outside the cell. Only after autolysis and cleavage of the bacteria, it can be released. Therefore, it is called endotoxin (Fig. 3-5).

Different Gram-negative bacteria have lipopolysaccharide of a similar structure. Therefore, all infections caused by endotoxin generate similar symptoms. These include the following.

The first symptom is fever. Endotoxins act on immunity cells in the body, causing the production of various immune factors. These immune factors act on the thermoregulatory center of the hypothalamus, prompting the body temperature to rise. The human body is highly sensitive to endotoxins. A very small amount (1–5 ng/kg body weight) of endotoxin can cause the body temperature to rise. The fever would last 4 h before subsiding.

The second is the leukocyte reaction. After endotoxin enters the body, the neutrophils in leukocytes move and adhere to the tissue capillaries, causing rapid decrease of neutrophils in the blood. In less than 1–2 h, the neutrophils induced by endotoxin will release factors that stimulate bone marrow to release neutrophils. These neutrophils then enter the blood stream, significantly increasing their numbers there.

The third is endotoxin shock. When a large amount of endotoxins enter the blood, they act on immune cells, prompting the organism to produce a series of biologically active substances such as interleukin. These substances can cause dysfunction of small blood vessels and microcirculation problems. Clinical studies show symptoms that include microcirculation failure, hypotension, hypoxia, and acidosis. In the end, the patient suffers from shock. This reaction is called endotoxin shock.

There have been painful encounters with endotoxin shock throughout history. In the 1940s when penicillin was first discovered, doctors found that penicillin was very effective on treating meningitis caused by *Neisseria meningitis*. Therefore, once a patient was identified, he was treated with selected penicillin. Meanwhile, according to general rules, the dosage increased with the severity of symptoms. However, accidents occurred. When treating patients with severe meningitis with large dosages of penicillin, many patients suffered from endotoxin shock and died. Later, through research and analyses, the cause for the problem was discovered. There is a large amount of pathogens inside these patients. Doctors tried to "flood" them with a large dosage, meaning to destroy the enemy in one swoop. Quick and thorough elimination of pathogens is not a wrong strategy. However, some doctors ignored the other side of the illness, which is that meningitis pathogens are Gram-negative bacteria, and their toxic substances are endotoxins. When killing all pathogens at once with penicillin, a large amount of endotoxins are also released. This causes endotoxin shock, and accelerates death of the patient. Current treatment still uses large dosage of effective antibiotics; in addition though, it is necessary to add hormone drugs to protect endotoxin-sensitive cells from reacting to endotoxin induction, thus helping the cells pass through the "shock." This is like using anesthetics during surgery so that the patient will not feel pain.

Human Immune System—Brave Attack on the "Invaders"

In our ever-changing lives, more than tens of millions of bacteria attempt to attack us every day. How does the human body protect itself? It is through rows of "defenses" and camps of "heavily equipped soldiers." These "defenses" and "soldiers" are able to not only keep invaders outside the body, but they can also attack and eliminate them. This is the human immune system; it has many functions. It can remove aging or damaged cells and control immunity.

It can also function as surveillance to detect and remove tumors. In the fight with pathogens, the immune system acts the role of defense. It identifies and removes pathogens, as well as produces various immune factors to neutralize toxins. This is how the organism protects itself from infectious diseases.

Human immunity to pathogens includes innate immunity and acquired immunity. Innate immunity is formed gradually in the long-term evolution of mankind. It is a born resistance that does not target specific pathogens. Therefore, it is also called nonspecific immunity. Innate immunity includes skin, mucous membrane, and the antimicrobial and bactericidal substances they produce. Examples include phagocyte, natural killer cells (NK cells), mast cells, and some of the effector molecules (Fig. 3-6).

Acquired immunity is acquired through the organism's contact with specific pathogens. It targets only these pathogens and is thus

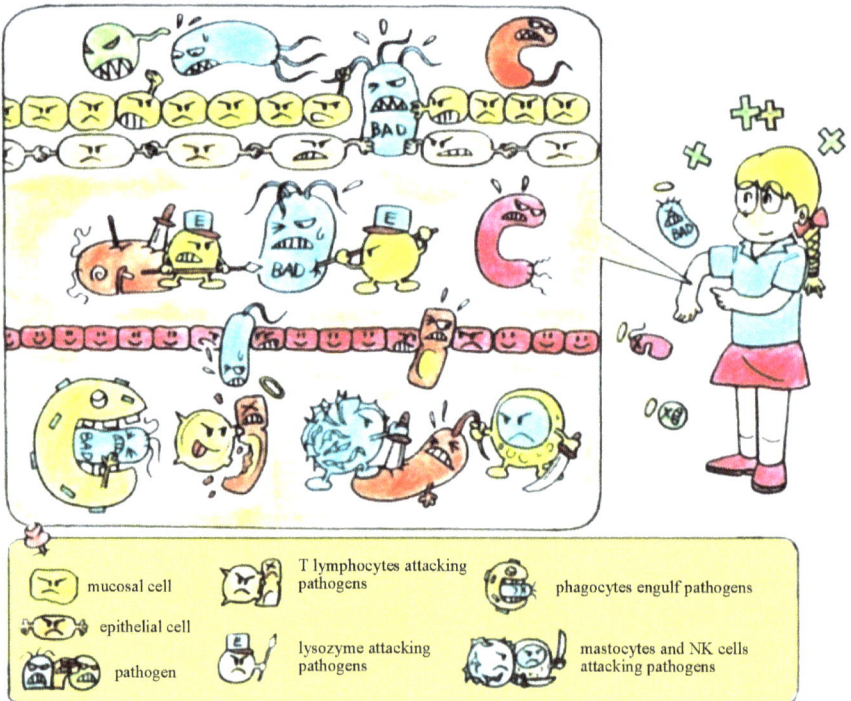

Figure 3-6: Human natural immunity.

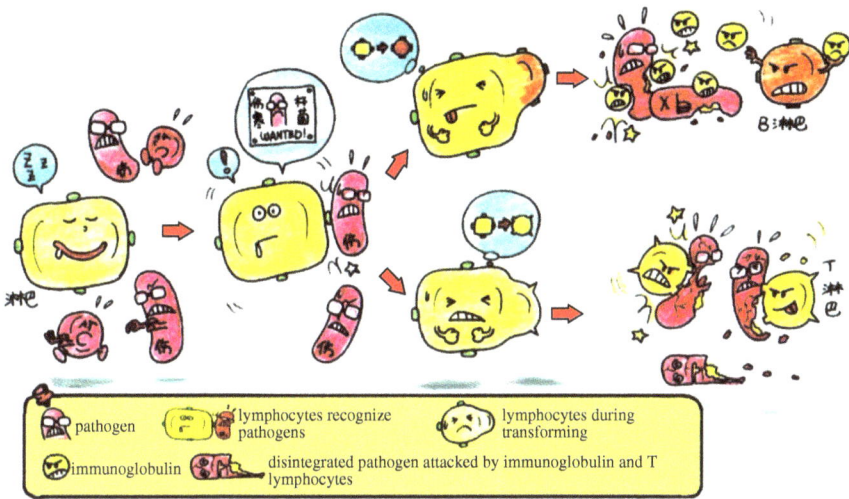

Figure 3-7: Human acquired immunity.

known as specific immunity. It functions through immune cells that can identify specific pathogens. Invading pathogens selectively stimulate specific lymphocytes that can recognize them. These lymphocytes proliferate and differentiate, forming effector cells (e.g., T Lymphocytes). These cells can directly devour or kill the immune functions of the cells. Other differentiated cells (e.g., B lymphocytes), can produce various types of immunoglobulins (antibodies). Releasing immunoglobulins into bodily fluids can have immunity effects. Although phagocytic cells can identify pathogens, they cannot identify specific pathogens. In specific immunity, their only function is antigen presenting instead of identification (Fig. 3-7).

"First line of defense" against bacteria invasion

A natural barrier hard to penetrate—skin and mucous

Membrane skin and mucous membrane are the natural barriers we use to protect ourselves against bacteria. The human skin is made of continuous and integrated scale-like epithelial cells. The cuticle on its outer surface is hard and impenetrable. This forms the first barrier to bacteria invasion. Meanwhile, sweat and sebaceous glands on the skin secrete lactic acid and fatty acid, which increases the acidity of the skin, thus inhibiting the entrance of pathogens. Our respiratory (bronchus and above), digestive, and genitourinary tracts (outer segment) are covered by mucous membranes. This is a weaker barrier. However, when pathogens enter the mucous membrane, organisms can get rid of pathogens through cilia movement, coughing, sneezing, and other mechanical methods. Meanwhile, mucous membranes secrete mucus, tears, saliva, and breast milk, which contain lysozyme, antimicrobial peptides, antibodies, and other natural antibacterial substances. In addition, tears, saliva, and urine have cleaning functions. Gastric acid in the stomach and spermine in semen have bactericidal effect.

There are large amounts of normal bacteria on human skin and mucous membrane. They not only take up lots of space so that there is no space for pathogens, but they can also inhibit pathogens. Some bacteria in the mouth, for example, can produce hydrogen peroxide, which kills diphtheria and meningococcus; *Escherichia coli* secretes colicin, which inhibits settlement and reproduction of pathogens.

Isn't it surprising that the thin layer of skin and mucous membrane play such a significant role?

Guard of central nervous system—blood–brain barrier

The blood–brain barrier is composed of soft meninges, outer wall of brain capillary, and astrocytes. Its features include close linkage between bacteria and weak endocytosis. It can prevent pathogens and other toxins in the blood from entering brain tissues, thereby protecting the central nervous system (Fig. 3-8).

It is important to note that the blood–brain barrier of babies is not fully developed; the blood–brain barrier of the elderly does not function fully due to low blood supply. Therefore, when infected, infants and the elderly are more susceptible to symptoms in the central nervous system, such as convulsions and coma.

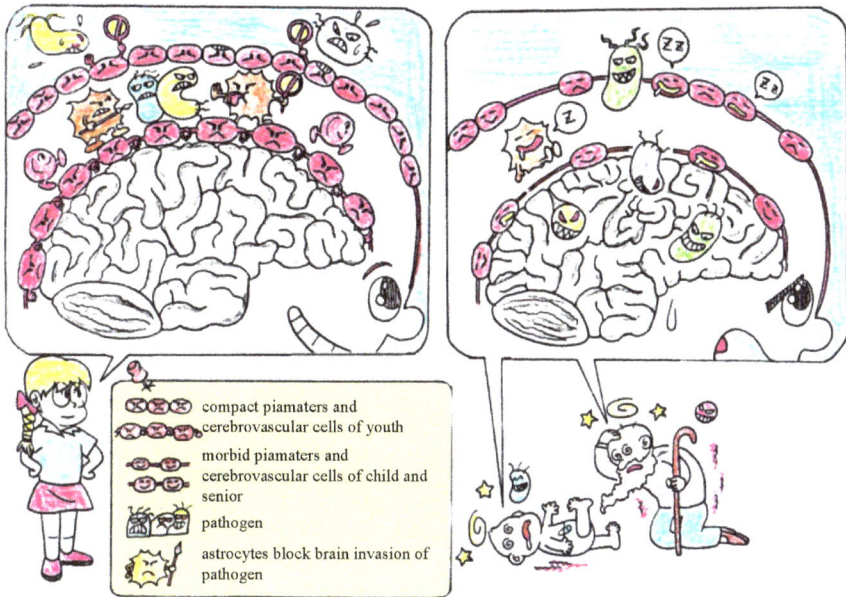

Figure 3-8: Blood–brain barrier.

Guardian of new lives—blood placental barrier

While still in their mothers" uteri, babies have another barrier, which is the blood placental barrier (Fig. 3-9). Blood placental barrier is composed mainly of maternal endometrium and fetal chorionic decidua. It does not interfere with substance exchange between mother and baby, but it can prevent bacteria and other microorganisms from entering the baby's body. Therefore, when blood placental barrier is fully mature, even when the mother is infected, the fetus remains unaffected by pathogens. However, during early pregnancy, especially the first 3 months, the blood placental barrier is not yet mature. Therefore, if the pregnant woman is infected during this time, pathogens will enter the fetus through the placenta, affecting the fetal development and even causing defects in the newborn. Similarly, medicine taken during early pregnancy can have the same effect because they can easily enter the fetus.

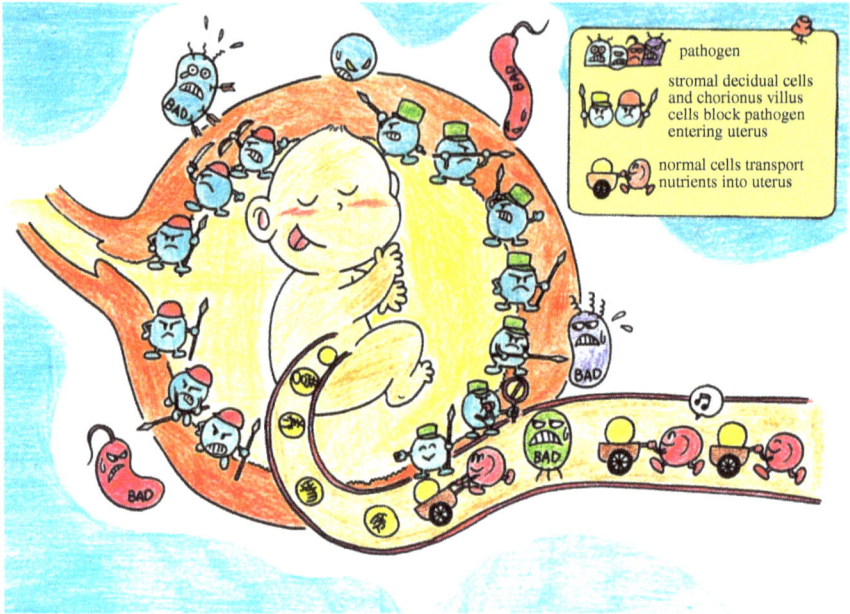

Figure 3-9: Blood placental barrier.

"Second line of defense" against bacteria invasion

Even after breaking through the first line of the nonspecific immune defense system, pathogens do not cause illnesses easily because they still have to penetrate the second line of the nonspecific immune defense system. This line of defense includes phagocytosis, antimicrobial substances in normal body fluids and tissues, as well as inflammation.

Bloody "body fight"—phagocytes" ability of phagocytosis

Based on size, phagocytes can be divided into small ones and big ones. Big phagocytes include monocytes and macrophages. Monocytes are a type of leukocytes. It is the biggest cell in blood. When monocytes leek out the vessel and enter tissues and organs, it will differentiate and develop into macrophages, a cell that has the strongest phagocytic ability in an organism. Small phagocytes are

Figure 3-10: Phagocyte devouring pathogens.

the neutrophils and eosinophils in the leukocytes. There are many receptors at the surface of phagocytes that can recognize and combine with pathogens (the receptor is a composing molecule of the cell surface. It can recognize and combine specifically with a substance called ligand). This activates the phagocytosis process. It also stimulates the lysosomes in phagocytes, thus killing the bacteria (Fig. 3-10). Lysosomes are small organelles inside the phagocyte surrounded by a single layer of lipoproteins. Inside, there is a series of acid hydrolases. It can decompose many substances. Therefore, lysosomes are called the "enzyme warehouse" or "digestive system" of the cell.

When entering the body, many pathogens secrete chemokines. Chemokines will attract a large number of phagocytes to penetrate capillary walls and move toward pathogen-affected parts. Most of the pathogens will be killed here; a few pathogens that are left enter the lymph nodes through lymphatic vessels and are killed by the phagocytes there. Even if a type of pathogen that is numerous, highly toxic, has not been stopped by lymph nodes, and has entered the blood stream and other organs; it can still be killed by the phagocytes in blood, liver, spleen, and bone marrow.

Some pyogenic bacteria, after being devoured, die after 5–10 min and decompose completely in 30–60 min. However, due to differences in the strength of the immune system among people and the type and toxicity of the pathogens, some pathogen, such as

Mycobacterium, has the function to inhibit formation of lysosomes or to resist the antimicrobial effect of lysosomes. Therefore, when such pathogen is devoured, it is not killed, but protected. It moves around with the phagocytes, causing the potential risk of disease.

Catching the slipping invaders—antimicrobial substances in body fluids and tissues

Normally, there are many types of antimicrobial substances in bodily fluids and tissues. They usually cannot kill pathogens directly, but they can cooperate with immune cells, antibodies (when pathogens invade the human body, immune cells produce immunoglobulin, which can act with pathogens to resist inflammation), or other defense factors, enabling them to better play their immunity role.

Complements are mainly produced by macrophages and liver cells. It is a group of nonspecific serum proteins in human serum. Normally, complements are inactive; however, it can be activated by the combination of the antigen (pathogen) and antibody. Because they play the role of supplying the antibody during antibody–antigen reaction, they are called complements. Activated complements can cause perforation of cell membrane, and the bacterium is dissolved. In addition, complements also have other physiological functions such as promoting phagocytosis, removing immune compounds, and promoting inflammation.

Apart from complements, other bactericidal or bacteriostatic substances include lysozyme, beta-lysine, histones, and interleukins. But they are less powerful, and they serve a complimentary role in the immune system. Figure 3-11 shows complements and other factors capturing slipping invaders.

Initial defensive response—inflammation

When the body is invaded by pathogens, the invaded part will demonstrate symptoms such as reddening, swelling, fever, pain, and dysfunction in the early stage. Later, it can suffer from abscess.

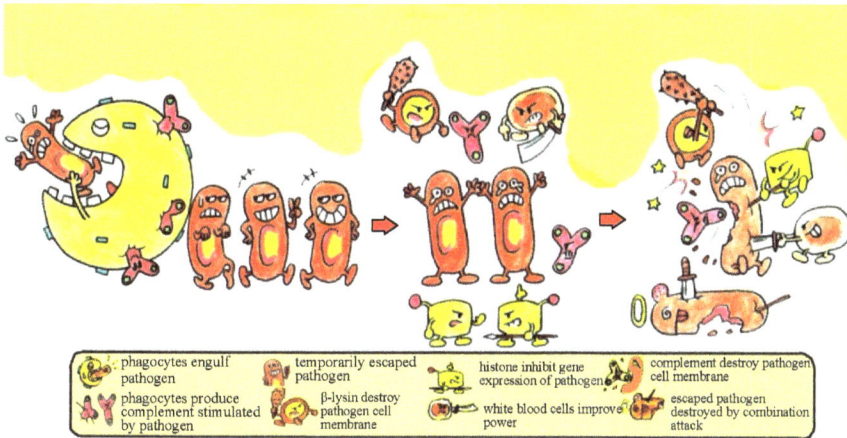

Figure 3-11: Capturing slipping invaders.

Actually, this is part of the local and systemic defense system of the human body toward pathogens—inflammation.

When pathogens enter the body, tissues and capillaries are stimulated and damaged. They send out "flares"—Histamine, serotonin and other inflammatory mediators—to activate an inflammatory response.

In early stage, capillaries at the site of the inflammation expand rapidly, the blood flow increases, and the permeability of capillary walls increases. Some soluble proteins can continuously leak out from the veins. Fluid accumulates at the site of inflammation. Meanwhile, the antimicrobial factors in the blood spread to these sites to take part in the battle.

As the battle continues, neutrophils in the inner wall of capillaries enter the camps of pathogens and devour them. But neutrophils can only live for 1–2 days. After their death, new signals are sent to aggregate phagocytes. Phagocytes move toward the site of inflammation and start the battle. But as they kill pathogens, they also damage neighboring tissue cells. During this process, bacterial toxins and other proteins are transmitted through blood to the thermoregulatory center at the hypothalamus, causing fever.

Figure 3-12: Defensive response—inflammation.

This is proactive defense because of the following reasons: (1) many phagocytes are driven to the site of inflammation; (2) acceleration of blood flow leads to the local concentration of antibodies and antimicrobial factors; (3) dead cells can release antimicrobial substances; (4) oxygen concentration drops at the center of inflammation, and lactic acid accumulates, which is advantageous to the inhibition of many pathogens; (5) high temperatures at the site of inflammation can reduce the reproduction rate of pathogens (Fig. 3-12). Therefore, inflammation in the organism can increase bactericidal capacities, reject bacteria invasion, and maintain stability of the body's internal environment. However, excessive inflammatory response can also cause severe damage to the body.

"Third line of defense" against bacteria invasion

Even if pathogens penetrate the first and second lines of defense, they still cannot cause disease easily. There is still the third line of defense—specific immunity. Specific immunity is the immunity that organisms develop against one or a specific type of pathogen. It varies from individuals to individuals as well as by different time periods in the same individual.

War against toxins — Antibody's neutralization of toxins

Pathogens are exogenous to the human body. Normally, some macromolecular components of pathogens (such as surface protein) can be identified by the immune system. These components are called antigens. When pathogens penetrate further into the human body, it will stimulate the lymphatic system to synthesize a globulin with specific immunity, which is also known as immunoglobulin, and this is the antibody.

The antibody is a large Y-shaped protein released by lymphoid B cells to identify and neutralize foreign substances (e.g., bacteria and viruses). It is used to tackle diseases, and it is the third line of defense for human health. Immune responses occur without notice. Therefore, even when we are exposed to an environment with bacteria and toxins, our body can still be healthy to some extent. When this last line of defense is broken, we start to get sick. Antibodies are Y-shaped; its two ends look like two pliers and can identify foreign substances. The identified foreign substance is called the antigen that corresponds to the antibody. The lower part of the Y is where the antibodies connect to the immune system. When the antibody has identified and combined with the antigen, the lower part of it will interact with other proteins, triggering a series of immunoreactions. This process is similar to someone who has discovered the enemy and while acting to inhibit enemy activities (combination of antibody and antigen), calls his men for help (stimulate immunoreactions). When other people (such as natural killer cell and phagocyte) come to help, the enemy can be defeated.

Antibody can be compared to the "soldier" emerging after pathogen (antigen) invasion of the body. It hunts the "enemy" bravely. After pathogens enter the organism for the first time, it is often after a period of latency that antibodies are produced.

Normally, the number of antibodies is not high, and it keeps decreasing. If, after a while, the same bacteria invade the organism again, the body will produce a large amount of antibodies rapidly and for a long time.

When exotoxin-producing pathogens such as diphtheria and tetanus bacillus invade, the body first produces antibodies to combine with the exotoxins, thus hindering the combination of the exotoxin with body cells; or the antibodies block the toxic parts of the toxin, hindering it from intoxicating the body. Next, phagocytes devour the exotoxin and the compound of the antibody. For those bacteria that do not produce toxin, the antibody combines with the antigen at the surface of these bacteria. The bacteria coagulate and thus lose the ability to move. The antibody can also prevent adhesion of pathogens, such as adhesion of oral streptococci to oral mucous membrane, adhesion of *Vibrio cholerae* to intestinal mucous membrane, or adhesion of pertussis coli to the respiratory mucous membrane. Pathogens attempting an invasion in the human body are identified by lymphocytes, which then produce large amounts of antibodies to neutralize the toxins produced by these pathogens, as shown in Fig. 3-13.

Figure 3-13:   Combination of antigen and antibody.

Siege—cell immunity

Some pathogens such as *Mycobacterium tuberculosis* and *Mycobacterium leprae* can stay for a long time in the human body after entering, and they reproduce. Antibodies cannot function directly on these invaders. At this time, cell immunity is needed to crush the invaders.

For example, when a person contracts *Mycobacterium tuberculosis* for the first time, although phagocytes can devour the pathogens, they cannot effectively kill them. Hiding in phagocytes, pathogens spread around the body, causing systemic infection. But during the infection, with the stimulation of pathogens, the phagocytes combined with pathogens can be identified by the lymphocytes in the blood and the surrounding lymphoid tissues. Lymphocytes then equip themselves with bactericidal equipment and activate lysosomes in the phagocytes. The pathogenic bacteria thus die of pyrolysis. In addition, lymphocytes release lymphokines, activate phagocyte, and dissolve the infected cells (Fig. 3-14). Figure 3-15 shows a variety of immune responses stimulated by antigens such as exogenous bacteria and toxins.

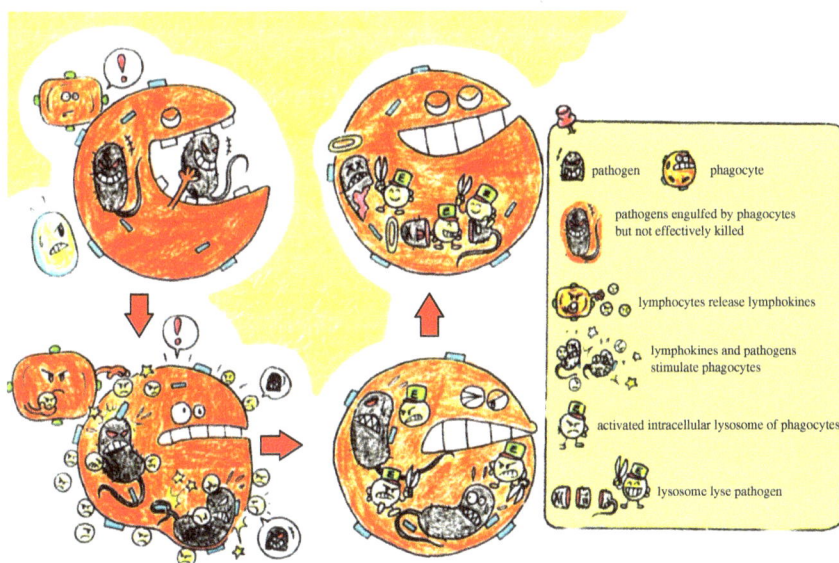

Figure 3-14:   Killing bacteria with cell immunity.

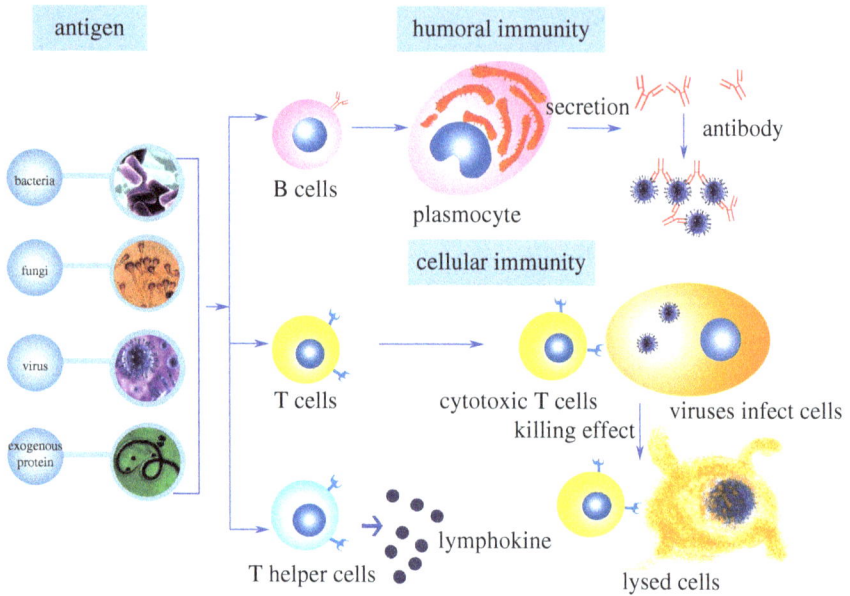

Figure 3-15: A variety of immune responses stimulated by antigens such as exogenous bacteria and toxins (from Introduction to Bioindustry by the same author).

End of the Immunity "War"

Some pathogens are highly toxic, such as *Yersinia pestis.* Invasion by a small amount can cause disease; invasion of only a few cells of this bacterium can cause infection. But for most pathogens, a certain threshold amount is required to cause an infection; invasion of a small amount will be removed by the defense system of the organism.

The site of pathogen invasion is also relevant to whether or not an infection would occur. Most pathogens are only infectious when invading through certain portals and residing in certain parts of the body. *Shigella,* for example, must enter through the mouth and reside in the colon in order to cause disease. For *Clostridium tetani* to be infectious, it should enter through wounds and reproduce at a local anaerobic environment in order to produce pathogenic exotoxins. When consumed along with food, it will not cause infections.

Once pathogens enter the body and attempt to damage cells and tissues, the organism will try to use various immunity functions to kill, neutralize, and repel pathogens and other toxins. During this "silent battle," the enemy and the body fight each other. The result of the battle is decided by the pathogenicity of pathogens, their numbers, their invasion routes, and the strength of the immunity system of the human body. As a result, some people suffer from disease, while others remain healthy.

Based on the comparison between pathogen toxicity and human immunity, there are only three possible situations that result.

The first is that the immune system achieves complete victory in the war. One situation is that pathogens enter the body, but they are immediately repelled or removed by the immune system, making them unable to grow, reproduce, or cause infections. Another situation is that invading pathogens are not very toxic or small in number. Their invasion causes slight damage to the body in the early stage, but seldom do they cause clinical symptoms. Meanwhile, the immunity of the organism is strong, and it eliminates the pathogens in a short time. This is also known as silent infection. After silent infections, the immunity will be enhanced.

The second result is that the two sides are evenly matched and engaged in a long-term confrontation. After pathogens enter the body, the immune system cannot completely eliminate or repel them. Rather, it restrains the pathogens in a certain part to inhibit it from reproducing and causing clinical symptoms. This situation is called carrier status. Carrier organisms are called carriers. They are a main source of some infectious diseases. If immunity falls again, the organism is likely to be ill.

The third result is that the enemy is way stronger than us, and pathogens gain the victory. If immunity is weak, or the invading pathogens are highly toxic and large in number, even though all defensive functions are activated, it is still difficult for the organism to prevent damage. Pathogens will soon reproduce on a large scale in the body and produce large amounts of toxins, damaging cells and tissues in the body. The organism will suffer from dysfunction and obvious clinical symptoms. This is infection (Fig. 3-16).

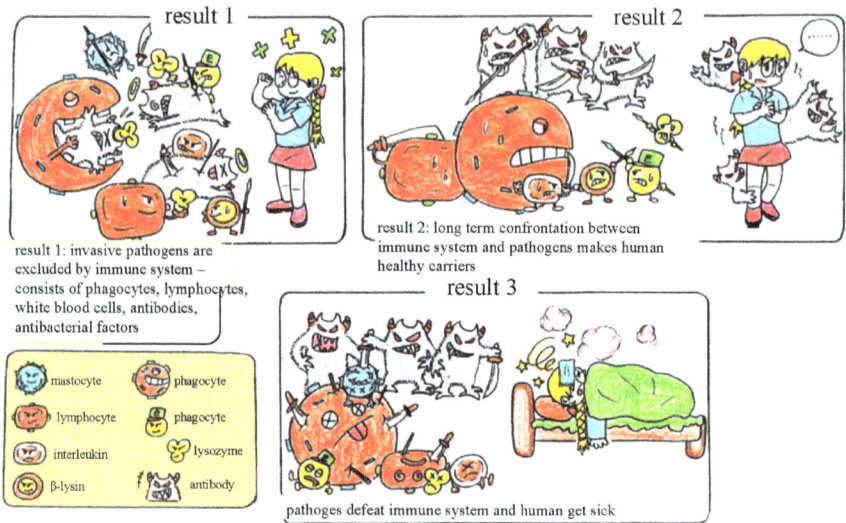

result 1

result 1: invasive pathogens are excluded by immune system – consists of phagocytes, lymphocytes, white blood cells, antibodies, antibacterial factors

mastocyte   phagocyte

lymphocyte   phagocyte

interleukin   lysozyme

β-lysin   antibody

result 2

result 2: long term confrontation between immune system and pathogens makes human healthy carriers

result 3

pathoges defeat immune system and human get sick

Figure 3-16: Outcome of immunity war.

Depending on the circumstances, infections are further grouped into acute infections and chronic infections, and local infections and systemic infections. Infection caused by *Vibrio cholerae* is an acute infection; its symptoms are abrupt short in duration. Tuberculosis is a chronic infection with a slow onset and long duration. Some infections are limited to one part, for example, bacterial infection of wounds. Some pathogens or the toxins they produce enter blood circulation and cause systemic spread of infection. Clinical symptoms include bacillemia, sepsis, toxemia, and pyemia.

Bacillemia is the situation where there are bacteria in the blood circulation, but as immunity is strong, the bacteria do not grow or reproduce, thus causing no obvious symptoms. Sepsis refers to the symptom of systemic infection caused by bacteria invasion in blood and their fast growth and reproduction. When infected, at first, chills are experienced. Later, the infected experiences a continuous fever in the range of 40–41 °C accompanied by sweating, headache, and nausea. Toxemia refers to persistent systemic fever accompanied by sweating, weak pulse, or shock, caused by bacterial toxin's entrance into blood circulation through the local infection. Because blood bacterial toxins can directly damage blood cells, it often causes anemia. It should be especially noted that, under situations such as severe injury, blood clots, and intestinal obstruction, although there is no bacterial infection, toxins are produced by a large scale and cause damage to tissues. This may also cause toxemia. Pyemia refers to systemic infection symptoms caused by new inflammation foci produced by bacteria in inflammatory foci "traveling" around with blood circulation. The symptoms are similar to that of sepsis; however, multiple inflammatory foci and many abscesses can be found on the body.

Chapter Four

A History of the Hard and Difficult War Against Bacteria

Antimicrobial Treatment during the Stone Age

In ancient times, people had little awareness of environmental health and possessed few methods to prevent and diagnose diseases. Diseases caused by bacterial infections such as cholera, tuberculosis, and dysentery not only affected the health of the patients, but also caused great damage to mankind.

In the stone ages, human prevention and treatment of bacteria-infected diseases were irrational. In ancient days, people did not know the existence of bacteria, let alone understanding the nature of bacterial infections. In the course of resisting various diseases, people learned to protect themselves; during difficult searches and accidental discoveries, people accumulated valuable experience for bacterial infection treatment. These methods and experiences appeared ineffective under the scrutiny of modern science. But at that time, these methods had indeed played important roles in prevention and treatment of bacterial infections; some are still used today or have inspired us in our battle with bacteria.

Why ancient humans were fond of silver and copper?

Why silver chopsticks?

Silver is a precious metal. It is shiny and bright, with delicate texture. It is often used for making rings, earrings, necklaces, collars, locks, bracelets, and hairpins, as well as tableware, wine ware, tea ware, and decorations. But, ancient humans did not fancy silver only for its grace and elegance. It was also because silver could be used for testing food poisoning; silverware turned black at contact with arsenic (arsenic trioxide).

Furthermore, silver also had long been discovered to have the ability of sterilization. There had been records of silver sterilization in ancient Egypt, Babylon, and ancient China. In 338 BC, when the Macedonians conquered Greece, they used silver films to cover their wounds so that the wounds healed quickly. Ancient Phoenicians used silverware for holding water, wine, and vinegar during their navigation to prevent spoilage. Ancient inhabitants of the Mediterranean put silver coins in wooden water barrels to prevent growth of microorganisms such as bacteria and algae. The *Compendium of Materia Medica* also contained such entry: "silver dregs can tranquillize the five organs, settle the mind, stop frights, and remove evil forces; taken in long term can rejuvenate the body and prolong life." Here, "evil forces" is referred to organisms invisible to the eyes at that time. Bacteria were one of them. During the Three Kingdoms period, for a quick recovery, soldiers often treated their wounds with silver. Because silver had antimicrobial capacities, the wounds usually healed five times faster than if treated with normal methods.

Miao people in China had always liked wearing silver bracelets and necklaces. Boys start wearing life-protecting bracelets at a very young age. After a while, a weak and thin boy will grow strong, his face is more red and his entire body is more energetic. Among common Miao folks, silver bracelets are believed to have the ability to strengthen and protect the body, an ability procured by everyone contributing part of their vivacity. In fact, this is also related to the antimicrobial features of silver.

In 1884, a German obstetrician put 1% silver nitrate into the eyes of newborns to prevent possible blindness caused by neonatal conjunctivitis. This reduced blindness rate of newborns from 10% to 0.2%. In 1893, scientist C. von Nageli conducted systemic researches, and he first reported silver's lethal effect on bacteria and other lower organisms and its potential to be used as disinfectant. Since then, human usage of silver entered the modern age. Silver silk is waved into gauze sponge to bandage wounds and release pain; silver compounds are used to treat burns in order to prevent infection and promote healing.

About 0.00002 g of silver ions in 1 L water will kill bacteria in the water. Disinfectants made of silver powder are 10 times more effective than chlorine disinfectants. Therefore, in water purifiers and mineral pots fabricated today, silver is often used as filters and sterilizing agents. Even under anhydrous conditions, silver still displays high bactericidal capacity.

In fact, high concentration of silver will cause slight irritation in upper respiratory tract or the eyes. Long-time exposure will cause argyria. However, the concentration level that is effective to bacteria is harmless to human. If a person takes 70–88 µg of silver per day, then 99% of the silver can easily be excreted.

Modern science discovers that silver does have strong capacity for sterilization and decomposing toxins. Silver can produce small amounts of silver ions in water. Positive silver ions can change the electrophysical properties of bacterial cells and damage their mechanical structure. The ions can also attract bacteria and disable the enzyme they relied on for breathing, thus killing the bacteria in a short time. Figure 4-1 shows the mechanism of silver ions killing bacteria.

Today, the antimicrobial ability of silver has attracted increasing attention again. Nano silver is the silver element in nanoscale size. Silver particles are thus thinner, their surface area larger, and their antimicrobial activity higher. The diameter of one particle of nano silver is usually around 25 nm. It is highly effective in inhibiting and killing dozens of pathogenic microorganisms such as *E. coli*, *Neisseria gonorrhoeae*, and *Chlamydia trachomatis*. Animal experiments show that nano silver antimicrobial powder will not be poisonous even when the dosage is several thousand times more than standard. Meanwhile, it can also repair damaged epithelial cells. When the bacteria are eliminated, nano silver ions are released from the cells and they form reactive oxygen in the air. Upon contact with new bacterial cells, reactive oxygen eliminate bacteria again, thus prolonging the bactericidal functions and making the nano silver ions

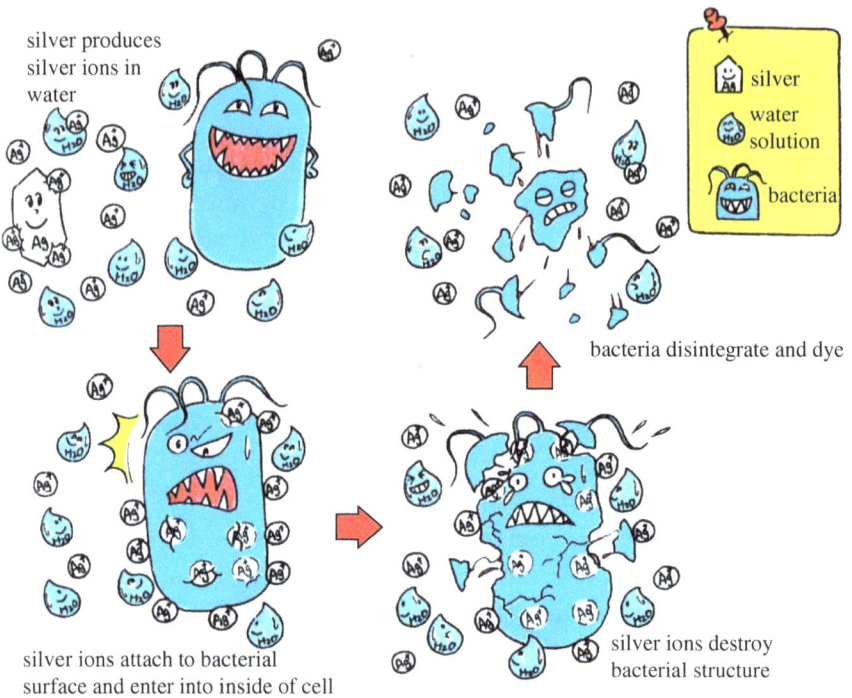

silver produces silver ions in water

silver
water solution
bacteria

bacteria disintegrate and dye

silver ions attach to bacterial surface and enter into inside of cell

silver ions destroy bacterial structure

Figure 4-1:   Mechanism of silver ions killing bacteria.

more effective. Nano silver is widely used. There are nano silver tea pots, nano silver antimicrobial sprays and fibers, textiles, and anti-bacterial construction materials containing nano silver. There are also new types of antibacterial gauze, first aid kits, and trauma hemo-static dressing material. Figure 4-2 shows the antimicrobial textiles made of nano silver.

Why make copper utensils?

Records of copper's antimicrobial capabilities appeared during 2600 BC to 2200 BC, in the medical book *Smith Papyrus*. The book recorded that copper could be used for disinfecting chest wounds and sterilize drinking water. There were many copper compounds for medical use. There were metallic copper cracks and chippings, as well as various forms of copper salt and copper oxide. The "green

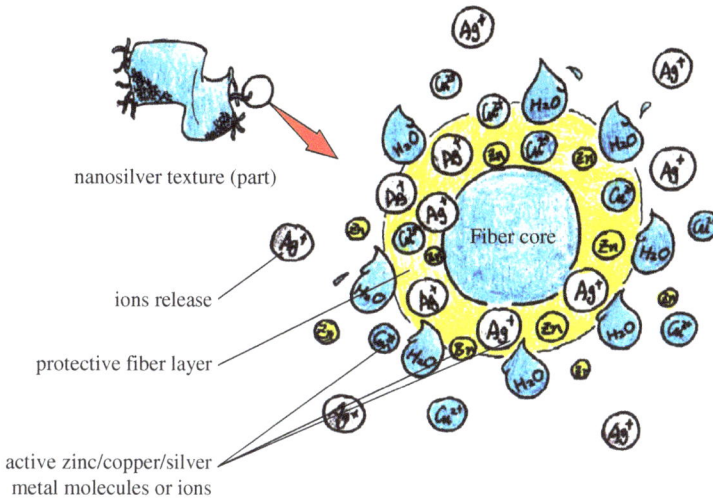

Figure 4-2: Antimicrobial textiles made of nano silver.

pigment" mentioned in the text probably referred to mineral mala-chite, a copper carbonate. It might have also referred to copper chloride produced by submerging copper in salt water.

To prevent wound infections, ancient Greeks used to put a kind of powder made from copper oxides and sulfates on the wound. Since there were plenty of copper on Kypros island, ancient Greeks could find the metal easily. It was because of this island that the Latin name for copper is Cyprus.

In *"Complete Works of Hippocrates,"* copper was recorded to have been used in the treatment of leg varices and other leg-related ulcers. In the book *Pliny,* there were many records of copper treatments. Black copper oxide mixed with honey can eliminate intestinal parasites; when diluted and dripped into the nose, it can keep the mind clear; when taken with honey, it can clear the intestines; it can also treat eye roughness, eye infections, blurred vision, and mouth ulcers; when blew into the ear, it could reduce symptoms of ear diseases. In ancient times, people covered wounds with copper for treatment and prevention of bacterial infection. Copper containers can keep drinking water clean. In addition to these usages, Verdigris and blue vitriol (copper sulfate) were also used for treating eye diseases such as bloodshot, infections, blurry vision, intraocular fat (trachoma), and cataracts. Scientists discovered, through research, two main mechanisms for copper's antibacterial function. One is through contact reaction; when copper ions come in contact and react with bacteria, it will damage the original components of the bacteria or cause dysfunction. The second is photocatalytic reaction. Under light, copper ions can catalyze and activate oxygen in water and air, producing hydroxyl radicals and active oxygen ions. These will damage the reproduction capacity of bacteria in a short time, thus killing the bacteria. Figure 4-3 is a demonstration of antimicrobial functions of copper.

Recently, Cupron, a US company, developed a new copper-based antimicrobial fiber named cupron. During the melting process, copper oxide powder is added to the mixture to obtain fiber with copper oxide particles. This fiber incorporated copper's ability to inhibit bacteria, viruses, and fungi, thus demonstrating fine antimicrobial and antiviral capacity. Since copper can stimulate regrowth of collagen in the skin, copper-based antimicrobial fiber can promote skin metabolism, thus making the skin smooth and healing damaged skin faster. It is highly valuable in functional medical dressings, antimicrobial textiles, and other fields. There have been many such fabrics made into underwear, socks, and pillowcases available on the market.

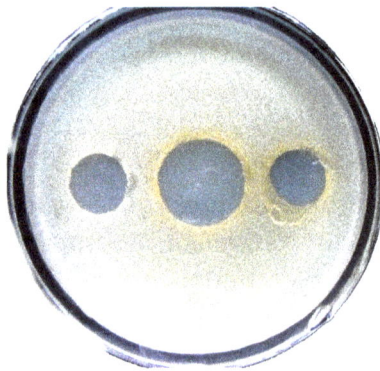

Figure 4-3: Antimicrobial functions of copper (bacteria growth was completely inhibited in places covered by copper).

Why hang wormwood for dragon boat festival?

The use of herbs has a long history, tracing back to ancient times. The knowledge, usage, and stories about medical herbs have been developing alongside evolution of human history.

The earliest evidence of usage of medical herbs appears in an archaeological site dating back 60,000 years. There are traces of used herbs in caves of prehistoric humans. These herbs are still used today for medicine. Along the Euphrates and Tigris 500 years ago and in Egypt, China, and India 4000 years ago, there had all been records or relics of laurel, caraway, mint, thyme, and other herbs. Papyrus in ancient Egypt and clay boards in Babylonian recorded names and functions of hundreds of herbs. In tombs of Egyptian pharaohs, people also discovered traces of herbs used for producing mummy preservatives.

Hippocrates of 5th century BC, who was revered as the father of Western medicine, was already familiar with the effects of many herbs. Back then, the plague broke out in Athens. Legend has it that he ordered Athenians to burn herbs on the street, inhibiting the spread of plague using its bactericidal effects.

Centuries ago, indigenous people of Australia called the Australian tea tree the "panacea" and used it for inflammation treatment. Modern medical research showed that the Australian tea tree has strong bactericidal abilities, a hundred times stronger than carbolic acid. It can be used in the treatment of Candida infections as well as others.

During the First World War, American government combined water with garlic juice and spread it on bandages. Application of these bandages on the wounded soldiers saved tens of thousands of lives.

In their long practices and experiences, the Chinese folk had accumulated rich and valuable medical heritages. The legend of Shennong testing hundreds of herbs was familiar to every household. *Shennong's Herbal Classic* from 200 BC recorded up to 200 herbs and their methods of use. *Taiping Shenghui Fang* and *ShengjiZonglu* of the Song Dynasty recorded a mouth spray named "fragrant ball." It was made of 15 ingredients: chicken tongue incense, patchouli, angelica, rosin, cassia bark, wood incense, nutmeg, Areca catechu, angelica, green cinnamon, cloves, and musk. The ingredients were ground to fine powders. Honey was added, and the powder was made into walnut-sized sugary balls. "Take a ball often and the throat is clear." Modern science shows that such "fragrant balls" often contain volatile aromatic oils. Some can resist bacteria and kill insects, some can smooth painful and itchy throat, and others can reduce inflammation. If consumed frequently, the balls can remove greasy feelings in the mouth, reduce oral odor, and strengthen the teeth. *Wubei Records* from the late Ming Dynasty recorded the recipe of a kind of Yam of Reassurance: wood incense, borax, nitre, liquorice, incense, realgar and cinnabar in equal amounts, with Mudingyang in half of the amount. Among them, wood incense can release spasm and resist bacteria; borax can detoxify and resist corrosion; nitre can detoxify and relieve swelling; incense can treat vomiting and stomach pain; liquorice can reduce pain and resist panic and anxiety; realgar can cure tetanus and epilepsy; cinnabar can cure mania, fright, poison swelling, and ulcer.

Bacillary dysentery is an acute intestinal infection caused by Shigella. As early as in *Treatise on Typhoid* by Zhang Zhongjing in Han Dynasty, it was mentioned that "the warmth of the sun causes disease that must be treated by releasing harmful substances; soup of kudzuvine root is the main treatment." Herbs that clear fevers, remove toxins, reduce body humidity, and release undesirable substances in the body are used for tackling bacillary dysentery.

Tuberculosis is a chronic lung infection caused by *Mycobacterium tuberculosis*. It can spread to the entire body, often breaking out when the immunity is low. It is called "persistent lung disease" in Chinese medicine. *Must Reads for Medical Persons* in the Ming Dynasty provided the following treatment: "all patients should not starve; weak ones need nutrition supplements, accompanied by benzoin and moschus." Treatment recorded in *Principles of Treatment* is as follows: "patients suffering from fever in the five core organs and are on the verge of lung disease should take heart-strengthening tuckahoe soup."

Chinese folks had also been using garlic as treatment. Army food supplies in ancient China included garlic for its bactericidal effects.

In written records, garlic from Jinzhou and Hubei were praised as the "bitter but effective medicine" and had been used for treating dysentery. During the Three Kingdoms era, Zhu Ge Liang was training an army in surrounding areas of Jinzhou. It was a hot summer and dysentery was spreading. Soldiers drank water with ground garlic. This treatment was quite effective. The famous "*Compendium of Materia Medica*" by Li Shizhen from Ming Dynasty recorded that garlic can "remove evil forces, kill toxins, treat diarrhea, dysentery, and wet cholera." Therefore, whenever epidemics broke out, people often put garlic in water to alleviate the "poison" in it.

Wormwood is one of the first plants Chinese folks came to know and use. According to experts, use of wormwood leaves had 2000 years of history in China. The first prescription book in China, *The Fifty-Two Prescriptions from Warring States*, recorded the effects and methods of use of wormwood. *Mencius* recorded, "seven-years of a disease require three-year treatment with wormwood." "*Zhuangzi*" also had similar record: "Yue people treated him with smoked wormwood." In *Wormwood Poet* (艾赋) by Kong Fanzhi, there was the record: "treat acute diseases with the magical wormwood, with the whole body covered in smoke." One can see that wormwood had already been an important medicine at that time. Wormwood is called "Ai" (艾) in Chinese. As to the origin of the name, *Compendium of Materia Medica* offered the following explanation: this herb can cure (乂) diseases; the longer the treatment the better; therefore it is called Ai (艾) after the character Yi (乂). Encyclopedia of Nature had the following record: "cut the ice to make it round; hold it to the sun, with Ai under its shadow, and fire is acquired," thus Ai is also called "Ice platform." Use of wormwood is not only through oral intake or acupuncture. Smoking wormwood is also used for treating and preventing diseases.

Li Shizhen and Li Yanwen in Ming Dynasty conducted thorough studies of wormwoods. Li Shizhen described in detail in the "Compendium of Materia Medica" form of wormwood leaves. He also discussed and corrected previous conceptions that wormwood leaves were cold and poisonous in nature. He also included 52 prescriptions with wormwood leaves. This is one of the herb books that provided the most wormwood prescriptions. In the "Compendium of Materia Medica," Li Shizhen pointed out the following: "Since the beginning, (wormwood leaves) from Qizhou has always been the best. It is an important medicine everywhere. It is called Qi Ai. Legend has it that wormwood from other places cannot penetrate wine barrels while Qi Ai can with much ease. This is special." This book is considered by later generations as the classic on Qi Ai, and has been cited in medical books. Qi Ai thus becomes renowned. Legend of Qi Ai by Li Yanwen is the first specialized book on wormwood leaves. It said wormwood "comes from the sunny side of mountains and are picked near the Dragon Boat Festival; they are used to cure diseases, their effect impressive."

During Dragon Boat Festivals in ancient times, people made "Ai figures" with wormwood leaves and used to hang them in the air: they cut the leaves into tiger shapes, which women hastened to wear, so as to repel evil forces. As "Jinchu Chronicle" recorded: "5th of May, pick wormwood leaves to make figures; hang the figures on the door to repel green gas." Modern researches find that wormwood leaves can significantly inhibit common inflammatory bacteria (*Pseudomonas aeruginosa*, *Escherichia coli*, *Staphylococcus aureus*, Alcaligenes). They can greatly reduce the number of bacteria on burnt wounds. They also possess considerable antimicrobial capacities against Proteus, diphtheria, typhoid, paratyphoid bacilli, and *Mycobacterium tuberculosis*.

In Jizhou, Hubei, where high-quality wormwood leaves are produced, there is still the popular saying, "three years with wormwood at home, never the doctor comes." At Dragon Boat Festival, the tradition of hanging and smoking wormwood leaves is still active. Some places preserve the habit of using wormwood leaves. Three days after birth, new born babies are washed in wormwood water; a few wormwood fibers are used for covering their bellybutton to prevent fever, stuffy nose, and other diseases. When an adult is infected with typhoid and cough, he can drink warm water boiled with wormwood leaves and leak, and wash his feet with wormwood water. Many plants contain antimicrobial substances such as Radix, leaves, *Artemisia annua*, and Agastache.

Many chemical medicines applied in contemporary clinics are first discovered in herbs, and then chemically synthesized for mass production. A typical example is quinine used for treating worm infection. In human history, malaria is a common and horrible disease. China's history of "western medicine" started with quinine. Back then, Emperor Kangxi was infected with malaria. The royal doctors could do nothing but witness the illness getting worse. By chance, there was a foreign missionary entering the palace. He recommended quinine, which instantaneously healed the emperor. Since then, medicines imported from the west are generally called "western medicine" in Chinese. Even later, when China can produce these medicines itself, the name is still widely used for distinguishing them from Chinese medicine.

The so-called western medicines are professionally known as chemical medicines. They refer to medicines composed of chemical monomers with relatively small molecular mass (usually smaller than 2000). Quinine is an effective chemical medicine to treat worm infection. It is a single compound extracted from the cinchona bark. Quinine has the strong capacity to kill malaria schizonts. About 1 g of quinine can be extracted from 10 kg of bark. Clinical investigations show that adult patients need 1–2 g of quinine every day for at least a continuous week to be cured. One can imagine the amount of bark required if one cures from malaria with cinchona-bark-boiled water! In addition, comparatively, there are usually no specific requirements on dosage when using herbs. The nonmedical ingredients in herbs are also not studied. Therefore, even water boiled with cinchona bark is effective; many people still cannot be cured due to insufficient dosage, or due to side effect of nonmedical components. Figure 4-4 demonstrates herbs' usage as medicine and shows the procedure for extracting chemical medicines from them.

Guasha and acupuncture

Guasha therapy

Guasha therapy, or scraping, is one of the most ancient treatments in Chinese folk medicine. When and by whom it was invented is still unknown. Doctor Wei Yilin in Yuan Dynasty wrote Effective Medical Prescriptions of the World （世医得效方）in 1337, which was the first to record Guasha therapy.

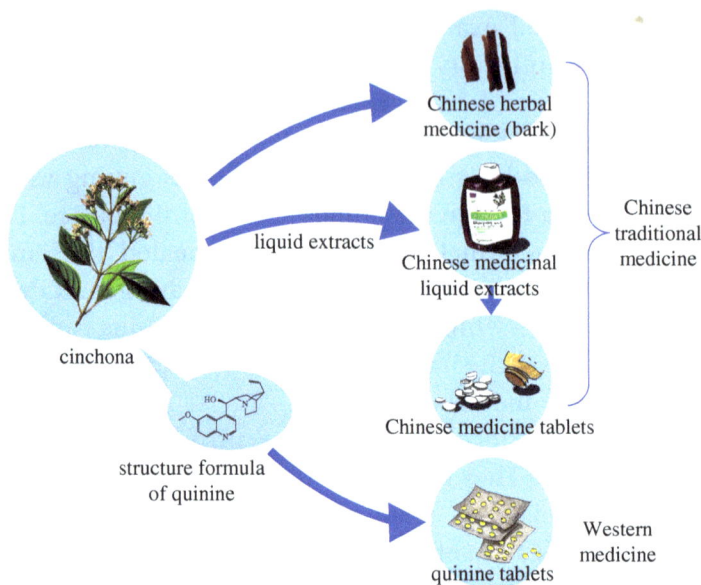

Figure 4-4: Direct usage of herbs as medicines or extracting chemical medicines from herbs (from Introduction to Bioindustry by the same author).

Guasha uses utensils to scrap human skin until capillaries of subcutaneous tissues break. Then, the blood flow out to form rice-like dots of red, purple, or black. This is called "sha." It takes several days before the "blood congestion" on skin subsides with autologous hemolysis. Guasha can remove from the body the undesirable factors from bacterial food poisoning, cholera, typhoid fever, scarlet fever, septicemia, and diphtheria, as well as pulmonary edma and syncope caused by accidental inhale of poisonous and smelly gas, thus healing diseases.

The release of "sha" is crucial to the effectiveness of Guasha. This artificial trauma is what Western medicinal professionals cannot understand, explain, or accept: as the objective is to heal, it is natural that pain and disease should be avoided. Then, why new wounds are added to the body? Chinese medicine offers an explanation: most of the diseases are caused by energy and blood. When virus invades the body, it blocks vessels, thus hindering the smooth circulation of energy and blood. The deeper the toxin penetrates, the more hindered the circulation and the more serious the disease. "Sha" is an excretion of toxic and undesirable substances. The release of "sha" means a way out for evil substances, thus achieving the objective to activate blood, adjust yin-yang balance, relax vessels, and improve energy and blood balance.

From the perspective of modern physiology, release of "sha" can produce a new stimulating hormone that enhances local metabolism. This can adjust nerve, endocrine, and immune systems, thus systemically adjusting the functions of all human organs.

What utensils are used for Guasha? Utensils for guasha are easy and convenient to use. Anything with smooth edge can be used, for example, combs, coins, spoons, and cup lids. For long-term use, buffalo horn is often used. This is because natural buffalo horns are nontoxic and nonirritating to human skin. Moreover, it does not produce undesirable chemical reactions. Furthermore, buffalo horn itself is a medicine. It has the function to diverge energy, activate blood, and nourish the body.

A layer of lubricant such as sesame oil or salad oil is applied to the skin before Guasha to prevent cutting the skin. "Guasha Blood Activating Gel" in professional use is a compound of natural plant oil and a dozen natural herbs. It can remove fevers, activate blood, open pores, repel toxins, release inflammation, and stop pain.

Acupuncture

Acupuncture originated from China. It has a long history. Acupuncture originated in the Neolithic Age. Back then, when people were ill or felt uncomfortable, they unconsciously patted and massaged with the hand, so as to release or eliminate symptoms. People often work in difficult environment to survive; therefore, they are often scratched or stun by sharp rocks, branches, or thorns. Sometime, there was bleeding. Occasionally, the symptoms of the diseases became less after bleeding. After repeated experiment, ancient people gradually realized that stimulating a part of the body or letting it to bleed could cure some diseases.

At first, traditional acupuncture used stone needles. Stone needle is a sharp piece of stone, often used to cut carbuncles or to release pus and blood. It can stimulate acupuncture points of the body and thus cure diseases. It is can be seen as one of the earliest medical tools. There are records in ancient books in China. Neijing, for example, recorded, "in the East land...all disease are ulcers and carbuncles. It is proper to cure these with stone needles." With the wider application of stone needles, people also invented bone needles and bamboo needles. When the ability for pottery was developed, ceramic needles were invented. With the development of metallurgy, people later invented copper needles, iron needles, silver needles, and gold needles. There is greater variety of types of needles. The range of disease that can be cured is also greater.

Moxibustion is the method to cure disease by lukewarm stimulation of certain parts of the body. Moxibustion originated from heating with fire. While around the fire, coldness was removed and people felt warm. Meanwhile, symptoms of old diseases were reduced or eliminated. Therefore, people wrapped heated stones or sand in animal hides or barks and warmed part of the body with it to release some pain. This is the original method of hot ironing. In practice, this method had been constantly improved. People used dried grass as fuel and applied lukewarm stimulation to local body to cure disease, thus developing moxibustion therapy. The fuel used in moxibustion was originally weeds or tree branches. Later, it was developed into coal moxibustion, bamboo chopsticks moxibustion, wormwood moxibustion, sulfur moxibustion, realgar moxibustion, and Juncuseffusus moxibustion. Chinese medical theories believe that the principle of acupuncture treatment is the balance of Yin and Yang, good and evil, as well as the clearing of meridians. As to antimicrobial features, acupuncture can cure dysentery. A clear understanding of the nature, seriousness, and acuteness of the disease to apply proper methods is required for treating dysentery with acupuncture.

Although acupuncture was originated in China, it had already been transmitted to Korea, Japan and other countries in 6th Century AD. In recent years, with the deepening of cultural exchange between China and the world, acupuncture is also spreading to all places of the world.

The Great Contribution of Immunological Prevention

As early as 1000 years ago, people discovered the phenomenon of immunity, and developed immunological prevention on this bases. Today, vaccine is one of the strongest weapons to prevent diseases. It protects millions of children and adults from infection of lethal diseases such as polio, tetanus, diphtheria, pertussis, yellow fever, Japanese encephalitis, measles, meningitis, and influenza. Vaccine has profoundly changed preventive medicine and has provided us more hope on disease prevention.

Precursor of immunological prevention—vaccination

Smallpox is one of the world's most infectious diseases. It is an acute and dangerous disease caused by variola virus. This disease regenerates fast and spread through air at a surprising speed. Death rate of smallpox is high, usually 25%, sometimes as high as 40%. If one is lucky enough to survive, he will suffer from permanent scarring or blindness. Smallpox outset is urgent. Patients suffer from headache and high fever, accompanied with nausea and vomiting. After 3–5 days, rash appears at the forehead, face, wrist, arm, trunk, and legs of the patient, which turns into herpes or impetigo in a few days. After 2–3 days of formation, impetigo gradually dries and forms thick scabs. After around a month, the scabs begin falling, leaving scars. In serious cases, patients suffer from sepsis, osteomyelitis, and other complications, which can lead to death.

People in ancient China discovered that survivors of smallpox could avoid the disease for a long time or even for their entire lives. Even some of them were infected with the disease again; however, the symptoms were light and not lethal. They were inspired by this phenomenon: some virus could be treated with virus, namely taking or contacting some toxic pathogens at the early stage of disease would give the organism specific resistance to this disease. Based on this, a method to prevent smallpox by inoculation of smallpox scabs was invented in China in the 11th century. Up to 17th Century, vaccination techniques had been greatly improved. It was not only widely used in China, but also exported to Japan, Korea, Russia, Turkey, Great Britain, and many other countries.

In the 18th century, China's smallpox inoculation technique reached Turkey via Russia. Mrs. Montagu, wife of the British ambassador in Turkey, witnessed locals inoculating children to prevent smallpox in Constantinople. She was impressed by the effectiveness of the method. Her brother died of smallpox; she herself had also suffered from smallpox. Therefore, she decided to vaccinate her son. Later, the boy was successfully protected from smallpox infection. When Mrs. Montagu returned to Britain in 1718, she became an active promoter of vaccination. Later on, this method became popular in Britain. British country doctor E. Jenner (Fig. 4-5) had been vaccinated when he was young. After finishing his education and returning to practice in the countryside, one of Jenner's responsibilities was smallpox vaccination. During practice, he became interested in a cow disease—cowpox. Cowpox is a mild form of smallpox. It was thus named because it was discovered in cows and other animals. Jenner observed an interesting phenomenon in his hometown: people who worked with animals seldom caught smallpox. The milking women believed that if they had been infected with cowpox, they would not be infected with smallpox. So they refused vaccination. This made Jenner surprised as well as puzzled. Inspired by the Chinese method, Jenner believed that the cowpox milking women had obtained a kind of resistance during work. This resistant protected the milking workers from smallpox. He made up his mind on an experiment to prove his hypothesis.

On 4 May 1796, with the consent of the father, Jenner carried out his experiment on an 8-year-old boy called Phipps. The boy had never had smallpox or cowpox. Jenner made two cuts on Phipps' arm and applied pox fluid from a milking women infected with cowpox to the wounds. Two days later, the boy felt slightly uncomfortable, accompanied with fever. However, not long after, the fever

Figure 4-5:   Great scientist, Jenner.

subsided. Only two scars of vaccination were left on the boy's arm. Would Phipps never be infected with smallpox again? Jenner could not yet draw the conclusion. His task now is to prove little Phipps obtained resistance to smallpox. Jenner extracted some pus from the scabs of smallpox patients, and planted it several times in Phipps' arm. A week passed, and Phipps did not catch smallpox. A month passed and Phipps was still healthy. Jenner's hanging heart was finally settled. Two years later, Jenner found a milking woman with cowpox and repeated the above experiment. The result proved once again that cowpox vaccination can prevent smallpox.

Jenner's method to prevent smallpox is called cowpox vaccination. This method soon spread around the world, and was later introduced to China by Portuguese merchants via Macao. Since cowpox is safer than human smallpox, Jenner's method was applied among Chinese folks to prevent smallpox. From then on, where there was cowpox vaccination, smallpox disappeared. In the 1970s, the world finally wiped out smallpox. Jenner' has not only protected human from smallpox; he was also the originator of natural resistance theory in immunology and the pioneer to apply this theory to promote health. To commemorate this great scientist, people established a statue in his hometown, with the words: salutations to the hero of mothers, children, and the people.

Less "poisonous" enemies are our friends

Although Jenner discovered cowpox vaccination, he did not offer a satisfactory explanation to why cowpox could prevent smallpox. Therefore, his invention was not further developed, but staggered for around 100 years. The person who really opened up perspectives for infectious disease immunity was French chemist and microbiologist Pasteur.

Cholera vaccine

In 1880, a horrible chicken cholera broke out in rural France. Chicken cholera is a fast-spreading epidemic. Its outset is fierce. Once one chicken is infected, the entire stock will die. Sometimes, people saw at one moment the chickens prowling around, and the next moment their legs shook, they fell to the ground, and died after a few blinks of their eyes.

Pasteur was determined to conquer this disease. In order to understand the cause of chicken cholera, Pasteur began with culturing chicken cholera bacteria as a breaking point. He was determined that chicken intestines were the best environment for the bacteria to recolonize, while chicken droppings were the media of infection. During his study, Pasteur successfully cultured chicken cholera bacteria in media made of chicken cartilage. When he applied one small drop of this pathogen to a healthy adult chicken, the chicken died immediately. This meant that the newly cultivated pathogen was as toxic as the pathogens in ill chickens. At this moment, an accident occurred in the laboratory. Due to the negligence of an assistant, the bacteria media was not planted timely into healthy chickens at designed time, but a few weeks later. Those chickens got sick at the beginning. But, later a phenomenon occurred that was previously unobserved: they did not die, but recovered, and were jumping around in the chicken house. When Pasteur heard of this amazing result from his assistant, he received a strong inspiration: what would happen when the recovered chickens were infected by the pathogen again? Pasteur infected these chickens with the cultivation media again. The result was exciting. The media could no longer do harm to the chicken.

Is it that the media left along for a while protected the chicken, so that they acquired immunity and would no longer be invaded by new pathogens? Pasteur's scientific insight made him realize that after a while, toxicity of newly cultivated pathogens was reduced or even eliminated. Meanwhile, Pasteur noticed that the longer pathogens were exposed to oxidization, the weaker their toxicity. When pathogens with weakened toxicity were introduced back into proper environments, such as human and animal bodies, they would reproduce again. However, under such circumstances, reproduced pathogens carried little toxicity. It was insufficient to cause disease, but could stimulate immune system to produce antibody, thus achieving immune effects. This is Pasteur's method of attenuation of virulent microorganisms and his principles of vaccination immunity—less virulent "enemies" are our friends (Fig. 4-6). Pasteur's attenuation vaccination can be compared with Jenner's cowpox vaccination. It also marked the discovery of a new weapon against infectious disease—immunological prevention, thus setting the foundation for immunology.

Anthrax vaccine

Anthrax is the notorious livestock killer. It is a serious infectious disease among horses, cows, and sheep. It can also infect human.

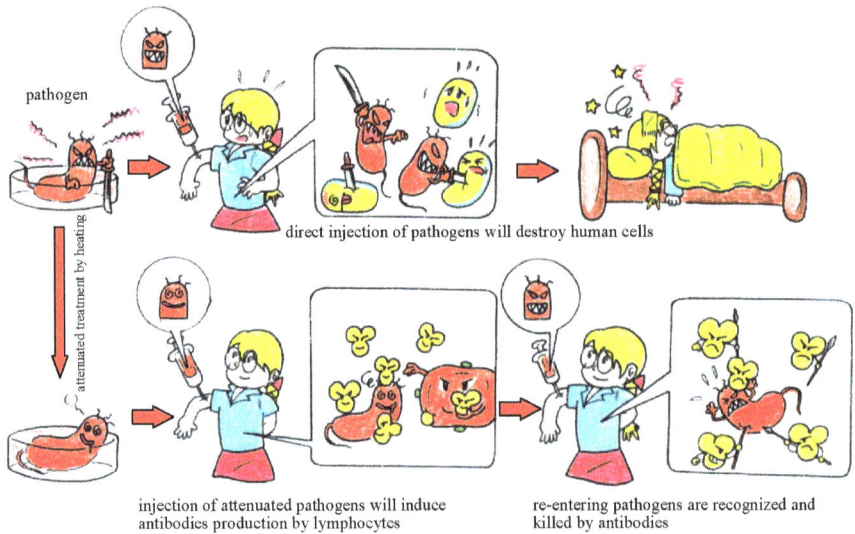

Figure 4-6: Less virulent "enemies" are our friends.

Shepherds and butchers are particularly susceptible to death from anthrax. There have been anthrax pandemics around the world. Before its outbreak, anthrax demonstrates no warnings. Animals almost cannot survive the disease. First, they have difficulties during walking, thus following back in the crowd. Their breath is heavy, and their bodies shake all over. Soon they fall on grass, and bloody stool stains the ground. Blood comes out of their mouths and noses. Before the shepherd can do anything, the beast is dead, line on the ground with limbs stretching towards the sky, and their stomachs swelling. If a knife is inserted into the still warm dead body, sticky and thick black blood will come out. An autopsy will surely find abnormally big spleens. Thus, anthrax is also known as leukemia.

In the early 1850s, a French doctor called Casimir-Joseph Davaine studied bovine anthrax. He discovered *Bacillus anthracis* in the blood of ill animals, and pointed out that it is the culprit of anthrax. In 1877, Pasteur started to systemically study anthrax. First, he extracted blood from animals died of anthrax, proving once again that *Bacillus anthracis* was the pathogen of anthrax. Secondly, he visited farm owners to understand the transmission of anthrax. Soon, Pasteur discovered that dead beasts were often buried in the pastures. Other beasts got sick after eating the grass at surrounding areas. This was because spores of *Bacillus anthracis* was spread everywhere in the pastures. But, the bodies were buried deep. How could the spores come to the surface and infect animals? Accidentally Pasteur discovered the role of earthworms. Earthworms were active underground. They stirred the mud and ate the body of ill animals. This way, spores of *Bacillus anthracis* were carried to surface ground. Therefore, Pasteur sent his assistants to collect the earthworms around the burying spots, and they indeed found spores in their bodies.

Pasteur instructed the herders against feeding the stock on grass that might hurt the mouth. This is because small cuts in the mouth were easy portals for *Bacillus anthracis* and its spores to invade. He also instructed against herding on burying spots. Burying spots must be carefully chosen. Ideally, these should be places difficult for earthworms to survive.

Pasteur published his research findings in February. Five months later, Professor Jean-Joseph Henri Toussaint announced his successful production of anthrax vaccine. When Toussaint released his method of acquiring vaccine through heating, Pasteur was outraged, because Toussaint's method killed *Bacilli anthracis*. Pasteur believed that the immunity reaction came from less-toxic pathogens, instead of from dead pathogens. Pasteur immediately proved with experiments that Toussaint's method could not produce effective vaccines. Meanwhile, he hastened his own development of anthrax vaccines. He knew that Toussaint had not given up, and was trying new possibilities.

During the experiments, Pasteur made another discovery. Some animals that were lucky to survive anthrax would not get sick when pathogens were injected again. This meant that they had acquired immunity. However, where to get these less-virulent pathogens that would not sicken the animals? Different from chicken cholera bacteria, *Bacillus anthracis* can produce spores to resist adverse living environments.

Through repeated experiments, Pasteur and his assistants found that constant culturing of *Bacillus anthracis* in environments of 45 °C would inhibit the formation of spores. Meanwhile, toxicity of the bacillus was reduced. Injection of these bacilli into animals would not cause disease and death.

In late February 1881, Pasteur released the news: he had successfully developed the new anthrax vaccine. Immediately, Pasteur conducted a public experiment in another farm. Some sheep were injected less-virulent *Bacillus anthracis*; others were not. Four weeks later, each sheep were given an injection of highly virulent *Bacillus anthracis*. After 48 h, sheep that were not previously injected all died; those who had been injected with less-virulent pathogens were still well and alive. Experts and journalists at the site were all very excited. The experiment was a complete success.

Rabies vaccine

In 1885, Pasteur completed the highest accomplishment of his life—vaccination of rabies vaccine. Rabies is caused by the rabies virus. Human can be infected when bitten by dogs and other animals who are carriers of the virus. Other warm-blooded animals such as cats and wolves can also spread infection. The most explicit symptoms of rabies include fear of light, fear of water, fear of motions, anxiety, and serious sense of fear. Fear of water is especially severe. While drinking, patients will suffer from pharyngeal muscle spasm and cannot swallow the water. Later, patients will not dare to drink water even when thirsty. Death rate of rabies is very high. Almost all cases died. It was believed that when he was little, Pasteur often heard screams on the street. People used hot iron to burn the wounds so as to "cure" people bitten by mad dogs. But, this was of no use.

In 1885, Pasteur made an attempt that would be remembered forever. As rabies virus was much smaller than normal virus, Pasteur could not isolate it and cultivate it in artificial media. So he used the spinal cord of live rabbits as media. At this time, a five-year-old boy died of rabies. After 24 h of the boy's death, Pasteur extracted saliva from the boy's mouth and diluted it with water. Then, the fluid was injected into five rabbits for observation. Soon, all rabbits died of rabies. Pasteur then extracted the spinal cord from one dead rabbit, and hung it in a microorganisms-proof bottle, waiting for the cord to dry and shrink. After 14 days, he took out the dried cord, grounded it, and added it to water to make a vaccine. The vaccine was then injected into the brain of a dog. The second day, sick spinal cord dried for 13 days was injected again. Thus, the injection continued for 14 days, each time with increased virulent. After a while, the dog was injected again with pathogens. The dog did not get sick. Rabies vaccine was successfully developed! Pasteur had finally found an effective way to cultivate rabies vaccine.

One July morning in the same year, a middle-aged woman with a worried face came to Pasteur's institute. She begged Pasteur to save her son who was just bitten by a dog. It turned out that, out of no reasons, the child was attacked and bitten outside the house by a mad dog that morning. His situation was serious. If not treated in time, he might have less than 5 days to live. Pasteur knew well that if vaccine was injected on this child, there were two possible outcomes: he was saved or he died sooner. After much consideration, he decided to vaccinate the child. Pasteur injected vaccine extracted from the rabbit spinal cord preserved for 15 days. The first day, he applied only a small dosage. In the next 10 days, he made 12 more injections, each time with a larger dosage. After a few days, the child miraculously recovered. Rabies vaccines for human use were thus born. This opened up a new era for rabies immunity and prevention. This is a great victory in human's resistance against the devil of rabies. Pasteur was revered as a "great scholar and the benefactor of mankind."

Make the "enemy" works for us

With cholera vaccine in 1879 and anthrax vaccine in 1881 as starting point, Pasteur's immune techniques had triggered great sensation. People soon realized that Pasteur's method could be used for preventing many other diseases. Therefore, a series of preventive vaccines were invented, for example, typhoid vaccine (1886), teta-nus vaccine(1890), pertussis vaccine (1926), diphtheria vaccine (1923), tetanus vaccine (1925), tuberculosis vaccine (1927), pneu-monia vaccine (1977), and influenza vaccine (1985).

Vaccine of bacteria is called bacterial vaccine. Bacterial vaccine is made of bacterial body. It has two types: dead bacterial vaccine and live bacterial vaccine. Dead bacterial vaccine includes pertussis vaccine, cholera vaccine, and typhoid vaccine. Bacteria are culti-vated in suitable media, and then killed, processed, and made into vaccine. This type of vaccine will not grow or regenerate after inocu-lation. Effectiveness of one-time injection is weak and short. Multiple injections are required before human body gains high and lasting immunity. Live bacterial vaccine is made from cultivated bacteria with no or little toxicity. These include Bacillus Calmette-Guérin and plague live bacterial vaccine. This type of vaccine can grow and reproduce in human body, producing long-lasting effects. Compared to dead bacterial vaccines, smaller dosage and less injections are required. It has better immunity effect and lasts for longer times. However, living bacteria requires refrigerating, thus it is valid for a shorter time. Its transportation and preservation are also less con-venient.

Apart from bacterial vaccine, another type of vaccine is called toxoid. It is made from bacterial exotoxins after detoxification. Common toxoid includes tetanus toxoid and diphtheria toxoid.

Bacterial vaccine and toxoid preserve pathogens' feature to stimulate human immune system, therefore when in contact with these harmless products, human body will automatically produce immunity. This kind of vaccination is called autoimmune. Vaccination is the most important means to prevent infectious diseases. It is also the most effective method. So far, there have been many vaccines successfully applied in immunological prevention.

Pertussis is an acute infection of respiratory tract among new born babies. Children under 5 years old are susceptible to infection, and they often suffer from severe paroxysmal cough, which affects breathing and eating. Pertussis is very dangerous to infants and children. Diphtheria is the respiratory tract infection caused by

Bacillus diphtheria. It produces white films in the nose and throat, which hinders breathing. It also secretes toxins that may lead to myocarditis or neuritis. Death rate is as high as 10%. *Clostridium tetani* invade through deep wounds. The oxygen-free environment is ideal for its reproduction. It releases large amounts of toxins, causing tetanus. Patients suffer from trismus, muscle contraction, and stiffness. DPT vaccine is made from a mixture of pertussis vaccine, diphtheria toxoid, and tetanus toxoid. It can prevent the three diseases at the same time. Bacterial toxins are injected into horse and other experiment animals repeatedly; when large amounts of specific antibodies are produced in the animals, horse serum is isolated; products made of concentrated and purified serum is called antitoxin, for example, tetanus antitoxin and diphtheria antitoxin. Antitoxin contains many antibodies. When injected into human body, the body gains immediate immunity without the need to produce antibody by itself. This is called passive immunity. Passive immunity is mainly used to treat diseases caused by bacterial exotoxins. The prevention period of these products are short, because they are soon excreted after injection. Therefore, they are usually used as temporary emergency prevention only.

Apart from using the above bacterial vaccines, toxoids, and antitoxins as vaccines; it is also possible to extract and purify effective immunity factors from bacteria using biochemical or physical methods to produce subunit vaccines. Examples include cholera toxin B subunit vaccine and *E. coli* pili subunit vaccines, as well as polysaccharide vaccine, which are composed of long-chain sugar molecules of bacterial capsule. With the development of life science in recent years, people have developed more vaccines with greater effectiveness.

Genetically engineered vaccine

It is also known as recombinant vaccines. It uses modern genetic engineering technology. The most identifiable, most specific, and most unique genes from pathogenic microorganisms that are most effective in stimulating immunity are recombined into cell expression systems of animals or microorganisms. This produces corresponding antigens such as proteins, DNA, peptides, and polysaccharides, which subsequently stimulates the production of antibodies. Once vaccinated, the human body will produce antibodies to resist invading pathogenic microorganisms. Since genetically engineered vaccines have high expression and few side effects, they are suitable for industrial production. They are cheaper than conventional plasma-derived vaccines, and soon they may become a type of vaccine product that all countries are eager to develop and produce.

DNA vaccine

DNA vaccines are also known as "naked" DNA vaccines, gene vaccines, or polynucleotide vaccine. It is a new field of study originated from research on genetic therapy in recent years.

Protein vaccine

Protein vaccine makes use of one kind of protein from pathogenic toxins or bacteria. For example, the capsular protein of *Streptococcus pneumoniae* can be modified so that it is less pathogenic to human bodies while its ability to stimulate production of antibodies is preserved. Therefore, when such vaccine is planted, organisms will not get ill, but can produce corresponding antibodies under stimulation to neutralize invading toxins or bacteria.

Peptide vaccine

Peptide vaccines are amino acid fragments produced in accordance with known or predicted part of the antigen's genes that can stimulate antibody production. A polypeptide made of several or several dozens of different amino acids are produced through chemical synthesis or genetic engineering. This polypeptide can stimulate production of antibodies but do not cause diseases.

Polysaccharide vaccine

Polysaccharide vaccine makes use of a certain type of polysaccharide from the pathogenic toxins or bacteria. Polysaccharides in bacterial capsules, for example, often display specific antigenic features during the pathogen's invasion of the organism. Through artificial modification of these polysaccharides, the toxicity of the pathogen is reduced, but can still stimulate production of antibodies. An example is active components, polysaccharide poly-phospholipid, or phospholipid ribose of *Bacillus influenza* vaccine. Polysaccharide of meningococcus can be obtained via chemical methods. After lyophilization, it can be used for preventing meningococcal meningitis.

The Discovery of Sulfonamides

In the early 20th century, pathogen was one of human's greatest enemies. Scientists had been searching for effective substance against pathogens. In 1910, German scientist Ehrlich synthesized Arsphenamine based on arsenic benzene compound, which was highly effective to syphilis. At that time people believed that they could finally conquer bacteria. However, their hope was dashed. Arsphenamine was not effective to bacteria. When people were helpless in front of bacterial diseases, another German scientist solved the problem. He invented the first antibacterial medicine in human history—Prontosil, opening the new era of chemical treatment of bacterial infection.

The scientist forced to give up Nobel Prize

Gerhard Johannes Paul Domagk (Fig. 4-7) was born in 1895 in Brandenburg, Germany. His father was a primary school teacher and his mother was a farming woman. Their family was very poor. In 1914, Domagk was accepted by Medical School of Kiel University with excellent grades. But, as soon as his study started, it was halted by the outbreak of the First World War. He volunteered in the army.

Figure 4-7: Domagk, originator of the era of chemical treatment to bacterial infection.

After the war in 1918, Domgk went back to Kiel University and obtained Doctorate of Medicine. After graduation, he worked in a large dye enterprise in Germany.

1927 was a turning point in Domagk's life. He was appointed as the director of Pathology and Bacteriology of the Bayer Institute. In 1932, Domagk discovered the significant compound—Prontosil. In 1939, as an acknowledgement of his significant contribution, Nobel Foundation decided to award him Nobel Prize in Physiology or Medicine. But then Germany was under Nazi rule. Hitler explicitly forbade Germany to accept Nobel Prizes. The Nazis force Domagk to sign for rejection of the Nobel Prize, and imprisoned him for 8 years.

Nobel Prize is only preserved for nominators for 1 year. Once the period expired, the prize will no longer be granted, but goes back to the Nobel Foundation. But, the medals and ceremonies to show respect for the winner can be preserved for a long time for the nominated. After the Second World War, Nobel Foundation specifically held a ceremony for Domagk. The Swedish King himself awarded the certificate and the Nobel Medal with his name to Domagk. On the ceremony, Domagk gave an enthusiastic speech titled "New Progress in Chemical Treatment of Bacterial Infections," which was warmly received by the audience.

Red dye—"Prontosil"

In the 1920s, medical field around the world witnessed an upsurge of synthesis of new organic drugs. After being nominated as laboratory director, Domagk devoted himself to exploring possibilities of medical application of certain dyes. He synthesized around 1000 types of azo dyes and tested their bactericidal capacity against *Streptococcus pyogenes* in test tubes. If the dye is bactericidal *in vitro,* he would proceed with experiment on mice. *Streptococcus pyogenes* was injected in mice so that they were infected with septicemia. Then, the bactericidal activity of the dye was tested again. However, the expected new antimicrobial drug did not emerge. Around Christmas Eve 1932, after numerous failures, a miracle finally happened. Domagk injected a non-antimicrobial red dye into infected white mice, and the mice gradually recovered. What is the red dye that cured the mice? In fact, it was already synthesized in 1908 and was called Prontosil. Because it can combine with wool protein rapidly and closely, it is often used to dye textiles. As there are some disinfectant factors in Prontosil, it was sporadically used for treating erysipelas and other diseases.

Domagk was both excited and calm at the discovery of Prontosil's medical values. He did not haste to publish articles. Instead, he only applied patent for it as "pesticide," because he felt further studies were required.

Prontosil is a red dye with various components. Exactly which component has the bactericidal effect? After numerous experiments, Domagk extracted a white powder from Prontosil, namely sulfonamides. Therefore, he did an experiment with sulfonamides on dogs. He injected hemolytic streptococcus into dogs. The jumping and alive dogs soon became ill, breathing heavily and unable to move. However, when sulfonamides was injected, the dogs were shaking their heads and tails in short time, and were gradually back to vitality.

This experiment proved that it was the sulfonamides extracted from Prontosil that was effective in killing hemolytic streptococcus. Out of prudence, Domagk did the experiment once again on rabbits and other animals, which all achieved expected results.

Any medicine is only convincing when clinically effective. How effective is sulfonamides to human bacterial infections?

The highest reward—saving daughter's life

When Domagk was preparing his clinical experiments, something happened to his family. One day, Domagk's beloved daughter Mary got a cut in the finger and was infected. Her fingers swelled with pain, and her entire body was in fever. Panicking, Domagk asked for the most famous local doctor, but nothing was helping. Mary's illness was not only controlled, but was deteriorating into sepsis, posing danger to her life. At this moment, Domagk thought: maybe I should know what bacteria caused my daughter's infection? He examined the fluid and blood from Mary's wound under the microscope, and discovered it was *Streptococcus pyogenes*. An idea flashed across his mind: sulfonamides. He was expecting to apply this new drug to human. The chance was here today. But, it has to be used on his beloved daughter.

Sulfonamides' successful application on animals did not testify its effectiveness on human. However, Domagk had no choice but to try. He injected sulfonamides solution into Mary's body in coma and observed his daughter with unfaltering attention, waiting for the miracle. Time passed in anxiety. Finally, Mary slowly opened her eyes. Domagk could not believe his own eyes. He steadied himself, looked into his daughter, stroked her forehead and said, "What a wonderful dream!" In Mary's eyes, which were larger than normal due to her fragility, the light of life was shown again. The daughter

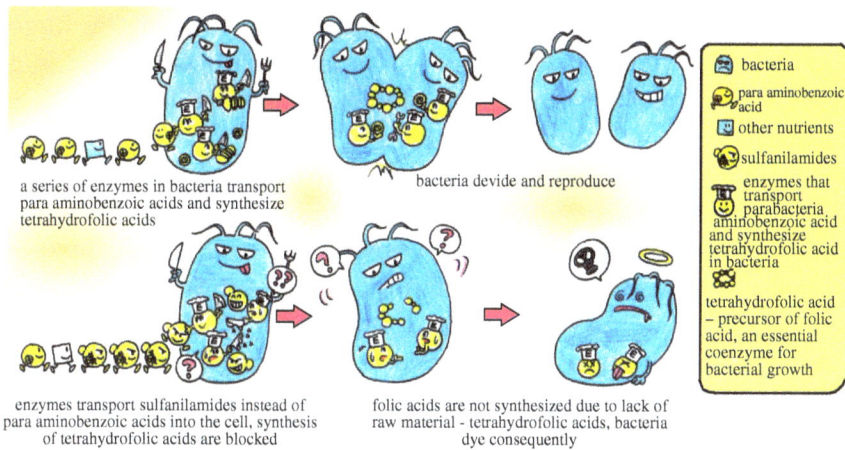

a series of enzymes in bacteria transport para aminobenzoic acids and synthesize tetrahydrofolic acids

bacteria devide and reproduce

enzymes transport sulfanilamides instead of para aminobenzoic acids into the cell, synthesis of tetrahydrofolic acids are blocked

folic acids are not synthesized due to lack of raw material - tetrahydrofolic acids, bacteria dye consequently

bacteria

para aminobenzoic acid

other nutrients

sulfanilamides

enzymes that transport parabacteria aminobenzoic acid and synthesize tetrahydrofolic acid in bacteria

tetrahydrofolic acid – precursor of folic acid, an essential coenzyme for bacterial growth

Figure 4-8: Antimicrobial effects of sulfonamides drugs.

is saved! The daughter in Domagk's arms was the first patient defeating streptococcal infection with sulfonamides in medical history. Domagk said with pride, "Curing my daughter is the highest reward for my invention." Discovery of sulfonamides caused great sensation in the entire world. Bayer Company produced the first sulfonamides drug in the world. Later, it has been produced in many other countries.

Soon, Pasteur Institute in France solved the mystery why Prontosil could only kill Streptococcus *in vivo*, but not *in vitro*. The reason was that when Prontosil entered the organism, it was decomposed to amino benzene sulfonamide (i.e., sulfonamides) through metabolic processes. Sulfonamides was structurally similar to p-aminobenzoic acid required for bacteria growth. It could be absorbed bacteria without providing nutrition, thus bacteria died of malnutrition. Figure 4-8 demonstrates the antimicrobial effects of sulfonamides drugs.

The persistent fighting power of sulfonamides family

Sulfonamides is the first antimicrobial drug. Today, it is still a commonly used drug for inflammation.

Sulfonamides drugs have many varieties, forming a huge "family." So far, more than 20 types are applied clinically. According to different types of infections, doctors choose different sulfonamides drugs. Systemic infection, for example, can be treated with sulfamethoxazole (SMZ), sulfisoxazole (SIZ), sulfadiazine (SD), cotrimoxazole (sulfamethoxazole and antibacterial synergist trimethoprim complexes), and bis-pyrimidine tablets (trimethoprim and sulfadiazine complex). Drugs used for intestinal infections include sulfa amidine (SG), succinylsulfathiazole (SST), phthalocyanine sulfonamide thiazole (PST), and phthalocyanine sulfonamide vinegar amine (PSA). Sulfacetamide (SA; SC-Na) is used for treating trachoma; silver sulfadiazine (SD-Ag) is used for preventing burn infections; sulfamylon (SML) is used in treating infection the after burn and large-scale trauma.

The Invincible Penicillin

The first sulfonamides drug was effective to purulent pharyngitis, meningitis, and gonorrhea. However, it was less effective to other bacterial diseases. In addition, many patients would suffer from side effects, even death. This urged scientists to find even less toxic new antimicrobial drugs suitable for wider application. The discovery of penicillin brought new hope to patients of bacterial infection. Since then, antimicrobial treatment entered a new age.

Opportunity favors the prepared

Doctor dedicated to bactericide research

Alexander Fleming was born in Lodge Field, Ayrshire, Scotland on 6 August 1881. He was the descendant of a poor Scottish farmer family. But, the fresh air, green fields, and clear brooks nurtured in him an unsophisticated nature. In the beauty of nature, Fleming learned at a young age to observe the nature carefully. When he grew up, a small unexpected legacy helped him into St. Mary's School of Medicine, fulfilling his dream for many years—becoming a doctor. Fleming was an assiduous student with outstanding curriculum. After graduation in 1909, he remained in the prevention and vaccination department of St. Mary.

Just when Fleming had all the ambitions to devote himself wholeheartedly to the field of immunological prevention, the First World War broke out. The War has changed the life tracks of almost everyone. Fleming joined the Royal Military Forces as a lieutenant. In a battle field, laboratory in Boulogne, France, Fleming participated in research and assisted in treating infectious disease of ally soldiers.

At the battlefield of blood and fire, the flesh body of human could not resist gunfire. Injuries became more and more. Although the injured were sent to Boulogne with the shortest delay, the poor sanitary conditions in battle fields meant that most injured were already infected when they arrived. Many died of bacterial infection of blood. At this point, even the best doctor could do nothing. Standing among those infected soldiers and looking at them suffering from pain at the verge of death, Fleming felt pain in his heart. He contemplated hard, and wished to develop a drug that could prevent wound infection. Gradually, this wish was consolidated in Fleming's mind.

In 1919, Fleming refused the invitation to practice in Scotland. He went back to St. Mary, and devoted himself to developing antimicrobial drugs. Fleming and his assistants targeted Staphylococcus. This was because it was a common and very dangerous pathogen. It was the cause of boils, carbuncles, tonsillitis, and fester in wound infections. Fleming and his assistants spent their entire days in their simple laboratory, inoculating and cultivating Staphylococcus in media, and then experimenting the effects of different drugs on Staphylococcus, in the hope of finding the ideal drug to kill it.

Fleming was a person of imagination and creativity. He never sticks to the rules in his work. In seemingly casual research, he kept making valuable breakthroughs. One day in 1922, Fleming, who had a cold at that time, was examining the bacteria in his petri dish. His nose was feeling uncomfortable. Suddenly, an idea struck him. He took some of his nose mucus and added it to a dish. A strange thing happened: almost all bacteria around the mucus where immediately damaged. Apparently, something in the mucus was lethal to bacteria. This accidental discovery attracted Fleming's attention. Since the substance secreted by mucus was an enzyme that could dissolve and eliminate bacteria, Fleming named it lysozyme. Fleming found lysozyme not only in human serum, tear, saliva, and milk, but also in white blood cells, egg white, radish juice, and other completely different substances. Unfortunately, the discovery of these taciturn bacteriologists did not produce much effect. However, the discovery of lysozyme pointed the way to Fleming for further research, and provided the foundation for his discovery of penicillin.

An accidental discovery

In an afternoon in September 1928, Fleming returned to his laboratory after a long vacation. He was examining the growth of his bacteria while chatting with colleagues. Suddenly, something caught his eye. He felt something was wrong. Without finishing his sentence, he rushed to observe one of the petri dishes. This dish was originally covered with golden Staphylococcus, but now they all turned bluish green. Since the dish had to be opened several times during the experiment, Fleming thought this must be because the Staphylococcus was contaminated. Strangely, however, wherever the cultures came into contact with the bluish green mold, the yellow Staphylococci were turning half-transparent. In the end, they were all lysed.

Undoubtedly, the mold eliminated the Staphylococci it came into contact with. Common scientists would not find this phenomenon significant, since by then it was already known that certain bacteria could inhibit the growth of others. But how could this unknown mold had such strong inhibition and lysis effect on Staphylococcus, one of the most dangerous pathogenic bacteria? Seen this way, the discovery became unusual. Sound scientific training made Fleming to realize that there might be something extraordinary. He wanted to know what exactly this powerful and mysterious mold was. He quickly scraped some mold from the dish and put it under the microscope. Through thick lenses, he finally discovered that the mold that dissolve and kill bacteria was *Penicillium*. Later, he made a series of culture media for *Penicillium*.

The result shows that this mold loved broth. With nutrition in the broth, it could grow into soft and fuzzy clumps in a few days. Spores were formed in another few days. Fungi clumps turned into dark green, and the media showed pale yellow. To his surprise, Fleming noticed that not only was *Penicillium* a powerful bactericide; even the yellow culture media possessed considerable bactericidal capacity. He so concluded that the real substance that killed the bacteria must be produced through metabolic processes of the mold. He called this substance penicillin. In the following 4 years, Fleming conducted comprehensive and specialized research on penicillin. He studied all kinds of molds in old clothes, old boots, vintage paintings, and other moldy things, as well as daily objects that grew molds such as cheese, jam, and other food. He collected all of them, put them in culture dishes, and observed if they possessed bactericidal capacities as *Penicillium* did. Finally, he found that penicillin was unique. Only penicillin could kill pathogens causing rotting wounds on the soldiers. Penicillin was one of the three discoveries of the Second World War. During this period, Time magazine published publicity figures of penicillin, with the caption, "Thanks to Penicillin, He will Come Home!" (Fig. 4-9).

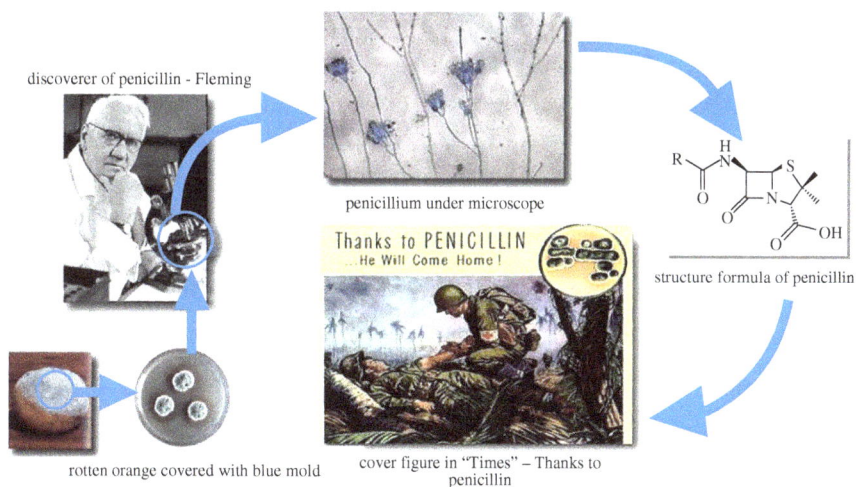

discoverer of penicillin - Fleming

penicillium under microscope

Thanks to PENICILLIN
...He Will Come Home !

structure formula of penicillin

rotten orange covered with blue mold

cover figure in "Times" – Thanks to penicillin

Figure 4-9: Discovery and application of penicillin (from Introduction to Bioindustry by the same author).

A series of experiments showed that *Penicillium* is a single plant fungus, not unlike molds in cheese or bread. But, penicillin can significantly inhibit and damage many pathogens that cause serious diseases. It also has high bactericidal capacity. Even diluted 1000 times, this capacity can still be preserved. Another advantage is that penicillin is little toxic or harmful to human and animals. Penicillin is the first effective antibiotics. Its magical power in curing infectious diseases brings good news to those who are on the line of death with disease. On the face of it, this significant medical achievement seemed quite accidental and incredible. Even Fleming himself called it a fortuitous opportunity in his report. In fact, as early as 1911, Richard Westling from Stockholm University mentioned penicillin in his PhD dissertation. It was proved to be the same penicillin discovered by Fleming. Unfortunately, Westling did not engage in further research, thus failed to find out its antimicrobial effects. Should people knew about this, they would have to admit that there was some certainty in the accidental—Fleming's personal characteristics and his research features: detailed observation, careful thinking, and keen judgment.

In 1945, at the commencement of Harvard University, Fleming gave a speech to 250,000 graduates, "one day in 1928, I had no intention to let *Penicillium chrysogenum* spores to fall on my medium; but the moment I saw the changes in the medium, I had no doubt that something extraordinary was going to happen." He earnestly advised Harvard Students, "ignore extraordinary phenomenon or event on no account. It might be a useless false alarm. But on the other hand, he might be the clue destiny provided you for great advances." He added, "when the mind is not well prepared, it will not see the hand of opportunity stretching in front of you." Great discoveries rely on careful work and a prepared mind.

Make publication

Fleming only had one kind of thin and diffuse nutrient solution that could be used in his early treatment experiments. He applied the solution to wounds, or used it to treat external infections of the eye, and he succeeded. So, Fleming did not hesitate. He truthfully published this result to the public. In May 1929, Fleming published his first several reports on penicillin on British Journal of Pathology and The Lancet. He enclosed original pictures of his first nutrition solutions, and wrote, "Penicillin might become an effective antimicrobial drug. It could be applied to skin or injected to microbial-infection areas that are sensitive to penicillin." Despite publication in two renowned journals, the reports did not cause much repercussion. At that time, the medical world had its attention on the promising sulfonamides. All other compounds seemed "trivial." Moreover, Fleming did not mention penicillin's success in clinical application. It successfully cured the eye infection of a miner, whose vision was preserved. In addition, penicillin also saved a baby with eye infection, who was infected due to the gonorrhea of his mother. Fleming used penicillin extracts to wash the baby's eyes and prevented him from blindness. When asked, many years later, why he did not mention these results, Fleming answered at that time, it was only crude extracts. The result was not sufficiently verified, thus could not be published. If he could have published these results earlier, wide recognition of penicillin might be advanced for several years.

As Fleming lacks related biochemical techniques, he could not extract sufficient penicillin for practical application. Meanwhile, Fleming discovered that penicillin producing fungus were highly unstable in his culture environments: 8 days into cultivation, the production of penicillin stopped. Gradually, even Fleming's colleagues lost faith in penicillin and gave up on the research. But, Fleming was convinced that penicillin was valuable. One day people would inevitably use its power to save lives. Therefore, he did not discard these bacteria easily. On the contrary, he continued culturing with much patience.

Florey and Chain—rediscoverer of penicillin

In Cambridge in the 1930s, Howard Walter Florey, an Australian professor of pathology, organized a large group of organic chemists, biochemists, pharmacologists, and bacteriologists to study lysozyme.

In 1935, under Nazi suppression, Russian-Jew Dr. Ernst Boris Chain escaped Germany and joined this group. While looking into the literature on antimicrobial substances, they had not only found Fleming's article on lysozyme, but also his paper on penicillin 10 years ago. The antimicrobial effect of penicillin stirred their interest, and the group immediately decided to put specific focus on penicillin research.

To proceed with the work, 250 pounds were needed to add some equipment and reagents. The insightful American Rockefeller Foundation provided the fund. In 1937, the group runs out of funding again, and their application was rejected by British Medical Research Council. Rockefeller Foundation stretched out a helping hand again, supporting the project for 5 years, with a total funding of 5000 dollars.

Florey and Chain discovered a type of *Penicillium* which was completely the same with Fleming's first discovery. They then proceeded without hesitation with experiment and isolation of penicillin. After countless sleepless nights, Chain finally succeeded in isolating the corn-starch-like yellow penicillin powder, and purified it for pharmaceutical use. This yellow powder was still antimicrobial when diluted 30 million times. It was nine times more effective than sulfonamide drugs at that time, and 1000 times more effective than the penicillin powder extracted by Fleming. In spring 1940, they conducted test tube experiments of 70–80 pathogens and various animal infection experiments. The results were quite satisfactory. Not only did penicillin show no explicit toxicity; it was also lethal to many pathogens. Figure 4-10 shows Florey, Chain, and their team who rediscovered penicillin.

Meeting of the three scientists

On August 1940, Chain and Florey published all their results on penicillin on The Lancet. This article greatly shocked one person, which was the discoverer of penicillin, Fleming. He had been paying close attention to developments on penicillin in the past 10 years. He was glad to see Florey and Chain's paper, because they had removed the long-standing doubt in his mind.

Figure 4-10:   Florey, Chain, and their team who rediscovered penicillin.

Fleming immediately set out to meet Chain and Florey in Oxford. This was a historical meeting. On the knowledge that Fleming was still alive, Chain and Florey were exhilarated. Without hesitation, Fleming gave the penicillin-producing bacterium he had been cultivating for years to Florey. With this bacterium, Florey and his team cultured penicillin with higher antimicrobial activity.

In recognition of their contribution to human kind, in 1945, Nobel Foundation awarded Nobel Prize of Medicine of that year to Fleming, Florey, and Chain. There was a small episode in the awarding ceremony: although some judges acknowledged that all the three of them were qualified for the prize, but they believed Fleming's contribution was greater than that of Florey and Chain; thus they suggested half of the price should go to Fleming, and the other half divided between Florey and Chain. In the end, however, the price was equally divided among the three of them, since most judges believed that without Florey and Chain's work, there would be no industrialization of penicillin or its significant contribution.

External environment is an indispensable catalyzes to success

Twists and turns in the fate of penicillin

The way of science to success is always long. After the rediscovery of penicillin, its fate was still full of twists and turns. British Medical Research Council and the University of Oxford not only refused to apply patent for penicillin, but also refused Chain's request for funding to build an experiment factory to further explore the industrial production of penicillin. Florey and his group tried their luck everywhere, hoping that British pharmaceutical companies would produce this promising new drug on large scale. Unfortunately, their request was ignored by most companies with the excuse of wartime difficulties.

Carrying penicillin producing strain, 2 g of penicillin, tiredness, and the remaining hope, Chain and Florey flew across the ocean to America.

Wartime needs

In 1939, the Second World War broke out. Eurasian continent was immersed in gunpowder. Many soldiers did not die of battle, but of wound infection. War required better antimicrobial drugs. The large-scale production of penicillin satisfied the pressing need. In 1941, the United States announced the priority status of penicillin in munitions manufacture, and listed documents concerning production and treatment of penicillin as "highly confidential."

In 1944, Anglo-American forces landed in Normandy, and began large-scale war against German Fascists. More and more soldiers were injured, and the need for antimicrobial drugs became more pressing. Although sulfonamides had played a significant role, it was not the ideal drug for treating heavily injured soldiers. At this time, the scarce penicillin filled the gaps of sulfonamides drugs, and demonstrated its capacity in healing the seriously injured. An army general heartily praised penicillin as a milestone in healing war wounds.

Development of fermentation industry

With the support of the military, penicillin proceeded on the way of industrial production.

Early method of penicillin production was solid surface culture, namely mixing material such as bran, soybean powder, and corn powder with water to make solid medium, which was heated and sterilized. Afterwards, it was mixed with penicillin bacteria and put into shallow dishes. The dish with the fermented substance was then put on an indoor shelf, kept under indoor temperature and humidity. The substance was constantly stirred for fermentation. After fermentation, penicillin was extracted from the solid medium with water, and was made into dry powder. The production process was similar to traditional methods used in rural areas for making soy sauce and vinegar. Although this primitive method could produce penicillin, there were still many problems.

To produce sufficient penicillin, large areas of media and culture room were needed. This took much space, and made control of temperature and humidity very difficult; meanwhile, workers worked in hot and humid environments under high work intensity. It was a hard job. More importantly, to insure ventilation during fermentation, the media was exposed to air. The microorganisms in air would cause much pollution, making pure fermentation impossible. Each time the result of fermentation was different. The process was difficult to control, and the quality of the product was hard to guarantee. There were still many other problems of the kind. Therefore, the production of penicillin using surface culture yielded low results. Yields were only 40 units/mL, and the yielding rate was 20%. Moreover, the product purity was 20%, and the cost was high. By today's pharmaceutical standards, these products could hardly be used as medicine. This urged people to explore new methods of production. In 1942, submerged fermentation was successfully developed. This method, contrary to solid surface culture, used liquid media for fermentation in certain containers. Liquid was relative to solid. In liquid media, water occupied 80–90%; whereas, in solid media, solid occupied 60–70% and water 30–40%.

Until 1942, the first batch of penicillin was produced in a factory in Illinois, the United States. But, the yield was very low. It was found that the *Penicillium notatum* discovered by Fleming did not yield much. Therefore, scientists started exploring other types of penicillin-producing bacterium. Finally, in a rotting melon in Peoria, a *Penicillium chrysogenum* was discovered. This mold reproduces rapidly, and its yielding was 200 times higher than *Penicillium notatum*. With this type, scientists used X-rays and ultraviolet mutagenesis to culture mutant types whose yielding was 1000 times higher than Fleming's original type.

Thus, in a short year, more than 20 American companies started mass production of penicillin with increased yielding. During the Second World War, it was exactly this amazing penicillin that saved hundreds of thousands of lives from death. During the First World War, 18% of American soldiers died of pneumonia. In the Second World War, however, the rate decreased to 1% with the help of penicillin.

The invincible penicillin

The discovery of penicillin is a milestone in the antimicrobial history of human kind. After large-scale application of penicillin, many formerly lethal diseases such as the incurable scarlet fever, purulent pharyngitis, diphtheria, gonorrhea, and various tuberculosis, septicemia, pneumonia, and typhoid have been effectively inhibited. The age where many diseases were incurable was no more. Average lifespan of human kind was extended.

The reason that penicillin was invincible to many bacterial infectious diseases was that it could prevent synthesis of peptidoglycan for cell wall, causing defects, water infiltration, and swelling of the cell wall. Cell dies of pyrolysis. Most bacteria are like eggs. Cell wall is the egg shell; it is like a thick wall, tightly wrapping all substances in the cell. Otherwise, these substances inside the cell will leak out. Thus, one can imagine that once the egg shell is broken, the entire egg is damaged; so it happens when the cell wall is damaged. Then, bacteria can no longer grow and reproduce. What will happen when bacteria get in contact with penicillin? From a macro perspective, the medium where bacteria are cultured show an apparent transparent circle (i.e., inhibition zone), as shown in Fig. 4-11(a). This zone is formed by molds living in the petri dish. The penicillin they produce inhibits the growth of surrounding bacteria. Under the role of penicillin, the cell wall of bacteria is damaged and the bacteria turned into a soft ball (i.e., protoplasts). Then, under osmotic pressure, the balls crack and the bacteria die, as demonstrated in Fig. 4-11 (b).

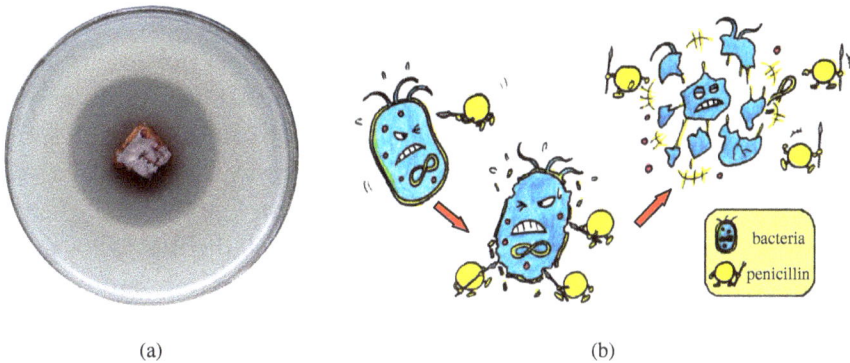

(a)   (b)

Figure 4-11:   Under the use of penicillin, cell wall breaks and the bacterium dies.

Penicillin has more than 70 years of clinical application. But, it remains an effective drug today. Especially, the emerging family of β-lactam antibiotics is the main drug against bacterial infections today. Apart from little allergic reactions, these drugs have almost no other toxic or side effects. This is because of the mechanism of these drugs is to inhibit formation of cell walls. As human cell has no cell walls, these drugs can exercise much selective virulence.

A great and ordinary man

Although before Fleming, some people had already observed some bacteria could inhibit bacterial growth, there had been no miracle. The discovery of penicillin undoubtedly makes Fleming a great man. However, from discovery to application, Fleming's penicillin was buried for 10 years. Apart from objective environment at the time, subjective conditions also restrained research on penicillin. After all, great men are also ordinary men.

Penicillin is hard to extract, and its activity is highly unstable. Under conditions at that time, extraction of penicillin remained a major challenge even for specialized biochemists. Fleming was a biologist. He was no all-rounder. Due to his lack of chemical knowledge, he could not extract penicillin from the culture media. Without purified penicillin, his research could not proceed, nor could it be applied to clinical practices.

Influenced by the restraints in science of that time, Fleming believed that test results on animals could not reflect possible results in human, or rather, it was unreliable to use animal test results as an instruction for human medical practices. This wrong thought dominated Fleming and his laboratory. The result was their insufficient confidence in clinical application of penicillin.

In addition, immunology was fast developing during 1920s and 1930s. Many infectious diseases could be effectively prevented through immunological methods; on the other hand, the new sulfonamides drugs demonstrated magic effects in treating many bacterial infections. Therefore, many scientists put the highlight of research on immunology or the development of sulfonamides drugs.

Apart from research interest of the scientists (Florey was the central figure for the rediscovery of penicillin), successful research, and application of penicillin relied on many objective conditions at the time: development in immunology slowed down; more and more soldiers were infected during the Second World War, while the formerly effective sulfonamides drugs were becoming less effective, namely bacteria were developing resistance against these drugs; large-scale microorganism culturing methods such as immersion (liquid) and sterile (pure) methods were maturing; and microbial mutation breeding techniques (to improve molds' ability to produce penicillin) were being applied and gained vast advancement in no time.

In fact, the combination of teamwork, research, and development of various disciplines has been indispensable. Were it not for the collaboration of scientists, purification and industrial production of penicillin will not become reality. With the joint effort of more than 200 scientists, the complex task is finally accomplished.

Significance of Streptomycin

Application of Penicillin has the epoch-making meaning in the history of antibacterial therapy, the situation that changed the way people fighting with the bacteria, which are in a passive state of affairs. However, penicillin even in small quantities is more precious than gold. The community demands urge scientists to seek and research new types of antibiotics. Furthermore, penicillin is not effective to fight mycobacterium, the pathogen of tuberculosis.

Tuberculosis is one of the most harmful diseases for human. In the 20th century, there are about 100 million people died from tuberculosis, including the worldwide famous writers who died from tuberculosis such as Chekhov, Lawrence, Lu Xun, and Orwell. Doctors worldwide have tried a variety of methods for treating tuberculosis, but still have not found a really effective way of treatment. Suffering from tuberculosis means death. Even after Koch discovered *Mycobacterium tuberculosis* in 1882, this situation has not changed for a long time. The magical effect of penicillin has brought a new hope for people, could we find the same effective antibiotic for treating tuberculosis?

As expected, after a few months of 1945 Nobel Prize Awards, in 1946, February 22, a professor at Rutgers University (USA) Selman A. Waksman announced that they found the second clinical used antibiotic in laboratory—Streptomycin, which has a specific effect against TB. A new era of fighting with tuberculosis has begun. Different from the discovery of penicillin, which is accidental, Streptomycin was discovered from a well-designed system after long-term study. Figure 4-12 Waksman (in the middle) isolating the antibiotic-producing bacteria from the soil.

Discovery of a magical medicine and a hostility of the scientific community

Waksman was born in 1888 in a Jewish family in Russia. He described his hometown once: It is located in a small village of the Siberian steppes, a cold village with fertile soil. In winter, it is covered with thick snow; whereas, in the summer, they harvest wheat, barley, rye, and oats. Fertile black soil nourishes all things, endlessly. Although in childhood Waksman lived in a village, this vast black land with no doubt had a huge impact to his future career choice. Unfortunately, the time was in the vortex of the Russian anti-Semitic movement,

Figure 4-12: Waksman (middle) in working.

because of which he could not complete his studies in Russia and fulfill his dreams.

When he was 22 years old, his family moved to the New Jersey, USA. The first few months he stayed at a relative's home. They had a farm. In that period, Waksman had enough time to get in touch with agriculture, such as animal husbandry, composting manure, and seed germination, which enhanced his interest in agronomy. His relatives asked Waksman to visit a Russian professor nearby in Rutgers College—Professor Lipman. Under the guidance of Professor Lipman, he gave up medical studies and entered the College of Agricultural studies.

Waksman is a soil microorganism scientist. He was interested in soil actinomycetes since undergraduate. In 1915, when he was still a Rutgers University undergraduate, he and his colleagues found Streptomyces—streptomycin was later isolated from this actinomycetes. People have noticed that tuberculosis in the soil will be quickly killed. In 1932, commissioned by the American Association against Tuberculosis, Waksman studied the phenomenon and found out that this is probably due to the role of microorganisms in the soil. In 1939, with funding from pharmaceutical giant Merck, Waksman with his students began to systematically examine whether antibacterial substances can be isolated from soil microbes. He would later name antibiotics as such substances.

Waksman's students reached up to 50. Their contribution to the project was enormous. In 1940, Waksman and colleague Woodruff (H.B. Woodruff) separated their first antibiotic—actinomycin, but its toxicity was too strong and not very helpful. In 1942, Waksman separated second antibiotic—streptozotocin. Streptozotocin has strong inhibitory activity against a lot of bacteria including *Mycobacterium tuberculosis*. But, still it is too toxic to the human body. During the Streptozotocin research, Waksman and his colleagues developed a series of test methods, it was extremely important for future discovery of Streptomycin.

Streptomycin was separated by a student of Waksman—Albert Schatz. In 1942, Schatz became a doctoral student of Waksman. Soon, Schatz was drafted into the army, to work in a military hospital. In June 1943, Schatz was discharged from the army due to illness, and returned to the Waksman's laboratory again to continue his study for PhD. The task assigned to Schatz is to discover new species of Streptomyces. He worked day and night for more than 3 months, in the basement that was transformed into laboratory. Then, Schatz separated two strains of Streptomyces: first was isolated from soil, second was isolated from a chicken's throat. These two strains are the same Streptomyces that Waksman discovered in 1915, but the difference is that they can inhibit the growth of *Mycobacterium tuberculosis* and other types of bacteria. According to Schatz, on 19 October 1943, he realized that he had discovered a new kind of antibiotic that is streptomycin. A few weeks later, after the confirmation that the toxicity of streptomycin is not too high, two doctors of Mayo clinic began to try it for the treatment of TB patients. The effect was surprisingly good. In 1944, the United States and Britain began a large-scale clinical trial, confirmed that the usage of streptomycin for treating tuberculosis is very effective. Subsequently, it was also confirmed that streptomycin is also effective to cure a variety of infectious diseases such as plague, cholera, and typhoid fever. At the same time, Waksman and his students continued to study different strains of Streptomyces and found that different strains have different production of streptomycin. Only four strains are able to massively produce streptomycin. In 1946, Schatz got his PhD and left the Rutgers University. Before leaving, upon Waksman's request, streptomycin patent was given to Rutgers University at free of charge. Schatz thought that no one would benefit from streptomycin patent, but Waksman had other ideas. In 1945, Waksman already realized that streptomycin will become an important drug, which will have a huge patent income. However, according to the agreement he signed with Merck company in 1939, Merck will own all patents of streptomycin. Waksman worried that Merck did not have enough support to meet the production needs of streptomycin; only if they allow other pharmaceutical companies to produce streptomycin, it will decrease the price of streptomycin. So, he asked Merck to cancel the 1939 agreement. Surprisingly, Merck agreed to transfer the patents to Rutgers University totally, only to keep the license to produce streptomycin. Rutgers University gave 20% of the patent income to Waksman.

Three years later, Schatz was informed that Waksman obtained personal income from streptomycin patent, and has reached $350,000. He was greatly dissatisfied and prosecuted Rutgers University and Waksman, and asked to share patent income. In December 1950, the cases got settled out of court. Rutgers University released a statement recognizing that Schatz is the codiscoverer of streptomycin. Under the settled agreement, Albert Schatz obtained $120,000 of foreign and 3% (approximately $15,000 per year) domestic patent income. Waksman received 10% of the patent income, and another 7% of revenue was shared by the other people who were participating in early research and development of streptomycin. Waksman voluntarily donated half of his patent income to set up Research Foundation to fund microbiology research.

Using popular words today, Albert Schatz broke the industry's hidden rules. Although he won the case, but cannot obtain a position in academia afterwards. He applied for more than 50 university faculty, but no one was willing to accept a "shyster," and he had to go to a small private agricultural college to teach. Although according to the law, Schatz was the codiscoverer of streptomycin, it was not admitted in academia. In October 1952, the Swedish Carolina School of Medicine announced that it will grant the Nobel Prize of Physiology or Medicine to Waksman as the discoverer of Streptomycin. Schatz disputed to Nobel Committee through his College of Agriculture to share the award, and asked a number of Nobel Prize laureates and other scientists for help. Few of them were willing to speak for him. On December 12 that year, the Nobel Prize of Physiology or Medicine was given only to Waksman. When Waksman describes the discovery of streptomycin in the acceptance speech, he did not mention Schatz instead of "we," only listed Schatz in the last acknowledgment. Waksman memoir was published in 1958. He did not talk about Schatz, but called him as "this PhD student."

Waksman thereafter continued to research antibiotics; together with his students, he found more than 20 antibiotics among which streptomycin and neomycin were the most successful. Waksman died in 1973, at the age of 85, leaving more than 500 papers and 20 books.

After this, Schatz never was allowed to work in any famous laboratories to do research. In 1960s, he could not find a job. So, he had to leave the United States to teach in the University of Chile. In 1969, he returned to the United States to be a faculty at Temple University. He was retired in 1980 and died in 2005 at the age of 84.

Schatz's contribution to streptomycin was almost forgotten. Only after retirement, people remembered about him. Thanks to the University of Sheffield's microbiologist Milton Wainwright. In 1980s, in order to write a book about antibiotics, Wainwright went to Rutgers University to check out the archives of the discovery process. For the first time, he learned about the contribution of Schatz. He then did a lot of investigation, and interviewed Schatz. Wainwright wrote several articles to introduce this event. Moreover, in 1990, he published a book describing the story of Schatz. Waksman had already died at this time. Thus, some professors at Rutgers University do not have to worry about embarrassing him. They also appealed for the return of his glory. By this, in 1994, when celebrating the 50th anniversary of the discovery of streptomycin, Rutgers University awarded a medal to Schatz.

Accompany with speaking out against injustice of ignoring Schatz's achievements was denouncing Waksman. For example, the British journal "Nature" in a commentary published in February 2002 had put this case as an example of injustice of the assignment of rights to the scientific discoveries. Schatz should be the real discoverer of streptomycin. In 2004, a writer, who was saved by streptomycin, published the "discovery of Dr. Schatz," together with Schatz. In the book, Waksman was portrayed as Schatz's scientific achievements appropriator, the man who took away all the glory of the discovery of streptomycin.

Did Waksman misappropriate the scientific research of Schatz? The best way to judge a scientific result is to see the published papers. In 1944, Waksman laboratory published the paper about the discovery of streptomycin. The first author is Schatz, the second author is E. Bugie, and Waksman is the last author. This author list is in full compliance with the practice of biology: Schatz was the one who mainly completed the experiment, so he was the first. Moreover, Waksman was the instructor, so he was the last. Waksman did not bury the visible contribution of Schatz in the paper. Their later dispute is because of the patent, which is not related to academic contribution.

Then, is it fair to award the Nobel price only to Waksman? Waksman or Schatz, who is the major contributor in the discovery of streptomycin? Streptomycin was not discovered by Schatz alone after several months of research, but was the result of years of systematic research in Waksman's laboratory. This should be mainly attributed to Waksman's designing. The work of Schatz was just a part of the plan. According to the research program and the experimental procedure, the discovery of streptomycin was just a matter of time. Schatz just performed the plan according to Waksman's research programs. With another graduate student, streptomycin will also be discovered. In fact, other students also discovered streptomycin from other strains later. The most significant contribution of Waksman was developing a systematic approach to the discovery of antibiotics, and this was also applied in other laboratories. Therefore, he was recognized as the father of antibiotics.

After all, the discovery of streptomycin should go to the leader of the project (advisor). The student who executes the project has the secondary contribution. In fact, this is the principle how Nobel Prize of Physiology or Medicine awarded not just applied only to the discovery of streptomycin but also other biological discoveries. The Nobel Prize is usually awarded to the leader of the experiment, but students rarely share. Schatz apparently knew this, so he always emphasized in persuading Waksman to study the anti-TB antibiotics. He was trying to make himself as a pilot project maker. But, this is not the truth, because before Schatz participation, Waksman laboratory already has tested the effect of antibiotics on *Mycobacterium tuberculosis*. However, the role of Schatz in the discovery of streptomycin should not be overlooked.

Interest of scientists in extensive research

Among other potential research, Waksman's early focus was aimed at separating sulfur bacteria. With the assistance of a colleague Stark, he successfully separated sulfur bacteria. This became the bright spot in a Waksman research career; also, he earned a reputation in the scientific community. Thereafter, lots of young researchers come to the lab bringing boundless vigor and vitality.

Waksman's research interests are vast. In the field of scientific research, he seemed to have unlimited energy. During 1920–1930, he gave lectures in many parts of the United States, Europe, and the Middle East, and committed to the systematic study of early marsh peat and compost. In many manufacturing companies and some other vitamins and enzymes companies, he worked as a senior technology consultant to provide technical guidance for these companies. In his efforts, microbial industry has gradually become a more biotechnological industry. In 1931, under Waksman's initiation, the United States established the Massachusetts Marine microbiology research center. Their main task was to cooperate with Navy to study antisepsis of the bottom of the ship. Afterwards, the study had a great significance to the US navy. This also earned Waksman reputations in the political field. In 1941, Waksman was elected as the chairman of the Strategic Committee bacteria and President of the American Association of bacteria.

The main achievement of Waksman was the discovery of streptomycin from *Streptomyces griseus.* In 1942, Waksman was the first who described antibiotic with a clear definition: antibiotics are chemical substances produced by microorganisms during metabolism and has inhibitory or even lethal activity against the growth of microorganisms. Waksman lived his life diligently, published several hundreds of scientific papers and reviews, 27 books, and one autobiography "My Life and microorganisms." In 1973, Waksman has completed his wonderful life, and quietly had a rest in a cemetery nearby Woods Hole. Accompany by many scientific pioneers, Waksman continued to receive respect from the visitors who come from around the world.

In the study of soil actinomycetes, Waksman established a scientific system and classification of microorganisms. Although he had not studied the active substances in the soil microbial to treat diseases, he is convinced that soil microbial species can affect the growth of fungi or bacteria.

In 1939, two events detoured the research of Waksman. First, the Second World War crashed the urgent need for new drugs able to control the infection and epidemic diseases; Second, one Waksman's student separated a substance that was effective in killing bacteria— tyrothricin. Although tyrothricin was very toxic to the cells, it reveled to Waksman a new theory: microbiology pioneers are very stressed by soil and other dirt that can lead to bacterial infection; but, in the soil, there are some microbes that are effective to inhibit bacterial growth.

Fleming observed the pathogen dish accidentally polluted by mold and fortunately discovered penicillin. But, the team of Waksman abandoned from traditional way to separate the antibacterial substance by lucky chance. In the streptozotocin research, they established a systematic approach to experiment and purposefully get antibiotics from microorganisms in the soil. They used different media to culture different microorganisms. Then, they observed inhibition zone around a single colony to target antibacterial activity of each single colony against a variety of pathogens. This is indeed a very complicated of precise work. There are thousands of microorganisms in the soil with different habits. Researchers must scrupulously separate microorganisms and then apply the nutritional requirements with different media and culture to obtain microbial metabolites. Finally, the activity of these metabolites on bacterial pathogens or other sterilization effectiveness was tested. Only a few microbial strains among thousands have antimicrobial activity, and the antimicrobial substances they produce may be too low to meet the production requirements, or some substances are too toxic to use. Thus, a new antibiotic was really picked from a thousand or a million, or even a used needle in a haystack to describe.

Inspired by the peculiar effect of penicillin and the incentives received by the discovery of streptozotocin, the team of Waksman worked tirelessly to find new antibiotics. In 1943, Waksman and his aides have isolated more than 10,000 microbes. Also, in this year, they isolated two gray actinomycetes (later named griseus), *in vitro* tests showed that the metabolites of gray actinomycetes could kill some pathogens.

Waksman understands that his laboratory no longer goes deeper on the research; so, he contacted a clinical laboratory to conduct a test on pigs. In 1944, the clinical laboratory spent an entire year to study pigs with TB infection by changing the dose to minimize the side effects. The results are exciting; the new drug has the treatment effect of tuberculosis and is no harmful to the animals. In 1945, a clinical trial of 33 patients also confirmed streptomycin is safe and effective. Waksman officially announced that a new antibiotic— streptomycin was born. Because of the experience for the production of penicillin, streptomycin soon went into mass production and rapidly became a "magic pill."

Application of streptomycin

Streptomycin has a strong antibacterial effect against *Mycobacterium tuberculosis*. Its discovery is a revolution in the treatment of tuberculosis. Since then, no special treatment but only resting and general supportive therapy was suggested in the era of TB therapy. Even today, it is still one of the first-line treatments for tuberculosis.

Although streptomycin is not effective against most Gram-positive bacterial infection than penicillin, it is very effective against most Gram-negative bacteria such as *E. coli*, aerogenes, *Klebsiella pneumoniae*, Shigella, Proteus, Brucella, and *Yersinia pestis*. It is a very good complement to penicillin, to treat Gram-negative bacterial infections. There is no cross-resistance between these two antibiotics. In case of drug-resistant strains, two antibiotics can be used interchangeably with each other.

Actinomycetes—the most Important Resource for Antibiotics Production

Another important contribution from Waksman is letting the people to know actinomycetes. This is the fundamental basis for the discovery of lots of new antibiotics later. Before, studies are focused on bacteria and fungi. However, Waksman made a new finding when studying soil microorganisms. There is a type of microorganism that looks like bacteria when cultured on agar medium, but after a period of time it looks like fungi because spores are formed by the colony. If cultured long enough, it cannot grow indefinitely like fungi. This is the wonderful actinomycetes. About 70% clinical used antibiotics are from this type of microorganism.

There are a lot of microorganisms in the soil. Researchers can isolate antibiotic-producing actinomycetes from soil by appropriate

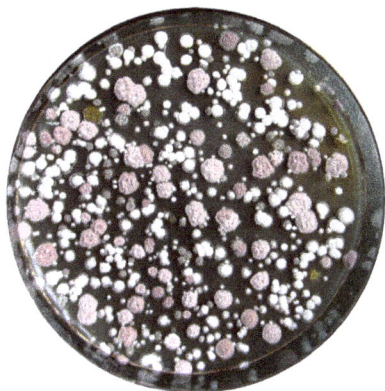

Figure 4-13: Varies actinomycetes from soil.

screening methods. Figure 4-13 shows the plate for isolating soil microorganisms. Those white colonies are actinomycetes. Further classification work is still needed to identify the type of actinomycetes. To date, scientists have extended the resource for antibiotics-producing microorganisms from soil to ocean, polar region, and plant endobacteria. Because of the biodiversity of microorganisms, there are more chances to obtain more and better antibiotics or other useful substances.

Different actinomycetes can form different colonies on agar plate. Even the same actinomycetes show different shape or color under different culture time or different conditions. Thus, identification of one type of strain (strain identification) has a stringent rule. Recently, gene sequencing can better identify the strain to avoid the frustration from different morphologies caused by different culture conditions.

Up to date, most of the antibiotics-producing actinomycetes are from soil. Actinomycetes are a big family among which Streptomyces produces the most antibiotics. The others include those rare actinomycetes such as Micromonospora, Actinomycesa lmadurae, and Actioplanes. The spores on top of the actinomycetes colony look like pearl necklaces or green and luxuriant grapes under the electronic microscope. Figure 4-14 shows the shape (colony) of some

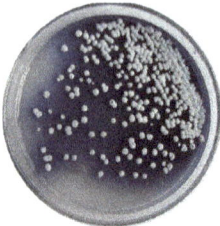

(a) *Streptomyces coelicolor* growing on solid medium

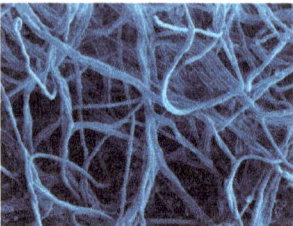

(b) spore chains of *S.coelicolor*

(c) *Streptomyces pristinaespiralis* growing on solid medium

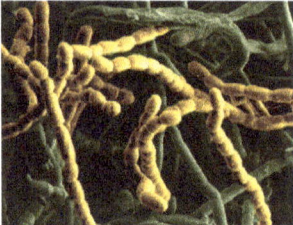

(d) spore chains of *S.pristinaespiralis*

(e) *Streptomyces roseosporus* growing on solid medium

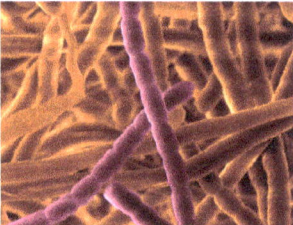

(f) spore chains of *S.roseosporus*

(g) *Streptomyces lavendulae* growing on solid medium

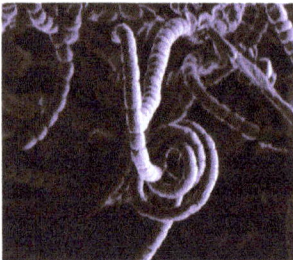

(h) spore chains of *S.lavendulae*

Figure 4-14: Shapes and spore chains of some members in actinomycetes family.

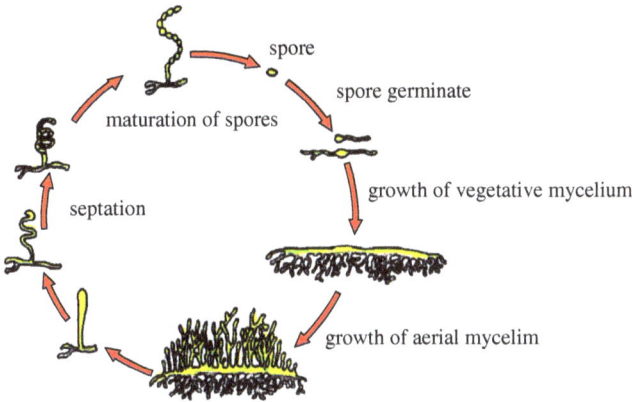

Figure 4-15:   Life cycle of actinomycetes growing on solid medium.

actinomycetes and their spore chains under the electronic micro-scope.

The life cycles of most of the actinomycetes are similar. Spores germinate to medium matrix and form small colonies. Then, it will keep growing towards the matrix to form substrate mycelium, and colonies facing up will grow towards the air to form aerial mycelium. Aerial mycelium can continuously grow in the air to form spores. Figure 4-15 shows the life cycle of actinomycetes grown on solid medium.

In liquid medium, most of the actinomycetes can form hair-like mycelium. They reproduce by budding. After the growth of mycelium, the antibiotics will start to secrete. Usually, they have a self-defense mechanism so they will not be killed by the metabolites produced themselves. Figure 4-16 shows the mycelium of actinomycetes under liquid culture.

The Golden era of Antibiotics Treatment

The exciting treatment effect of penicillin and streptomycin in clinic triggered the research from all over the world. Moreover, it provided the confidence for scientists to invent new antibiotics. Giant pharmaceutical companies invested a lot of efforts on antibiotics R&D. This caused a world competition on the discovery of new antibiotics.

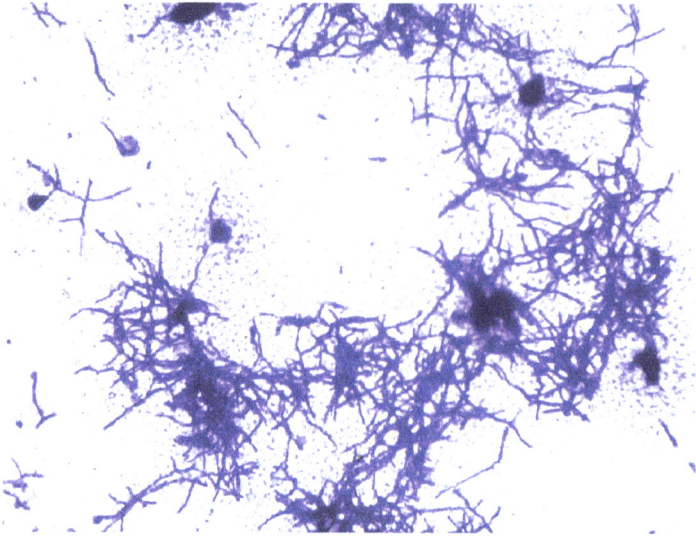

Figure 4-16: Mycelium of normal actinomycetes in liquid culture.

During this time, these studies are well planned, more targeted, and more systemic, and the production went on industrialization. A large scale of antibiotics industry was established while a lot of antibiotics were used in clinics. This is the golden era of antibiotics.

The discovery and application of a series of new antibiotics made bacterial infection to be effectively controlled. It is also the golden era of antibacterial treatment for humans.

New antibacterial weapons emerge endlessly

A series of systemic methods established by Waksman for finding antibiotics from microorganisms put the discovery from experienced, perceptual to rational, scientific stage. Using these methods, Waksman found 20 nature antibiotics in his entire life, most of which were found from actinomycetes. Actinomycetes are the most important resource for the production of antibiotics.

During this golden era period, researchers used Waksman's methods for collecting samples from sewer gutter, trash stack, or fertile field, and then screened the strains, tested the antibacterial activities. Polymycin (1947), chloramphenical (1947), aureomycin (1948), neomycin (1949), terramycin (1950), erythromycin (1952), carbomycin (1952), nystatin (1952), tetracycline (1952), trichomycin (1952), oleandomycin (1954), seromycin (1955), novobiocin (1955), kanamycin (1958), rifamycin (1959), lincomycin (1962), clindamycin (1967) were gradually discovered within 20 years. It is a grand occasion of antibiotics. Although we are now in the 21st century, most of the clinical used antibiotics are found in 1950s or 1960s.

The emergence of different types of antibiotics made possible of curing almost all bacterial infection and significantly increased human's life.

Double the power of the weapons

Reorganization of penicillin

Along with the wide use of penicillin, the resistant strains have become more obvious in clinic. Meanwhile, to overcome the allergic reaction and oral availability of penicillin, the medicinal chemists tried to reorganize the structure of penicillin to look for more powerful derivatives. This opens up the era of semisynthetic antibiotics from 1960s.

In 1959, the scientists from the British Beecham laboratory isolated and purified 6-aminopenicillanic acid, the core structure of penicillin, for the first time from penicillin fermentation broth, and synthesized a series of new penicillin by acylation of 6-aminopenicillanic acid. This is the start of a new era of the chemical modification of existing antibiotics. This also opened up a new path for the development of antibiotics. Over the next few years, people use the core structure of penicillin to produce a huge penicillin family by chemical modification (e.g., different side chains). This included different characteristics of semisynthetic antibiotics such as pheneticillin (phenoxyethanol penicillin), oxacillin, amoxicillin, ampicillin, methicillin, and piperacillin.

Structural modification of cephalosporins

In 1961, the scientist Abraham found cephalosporin C from Acremonium metabolites. In light of the development of semisynthetic penicillin, also due to the development of synthetic chemistry and technical improvement, scientists hydrolyzed cephalosporin C followed by ligation of different side chains for producing many highly active semisynthetic cephalosporins. Since the core structure of cephalosporin has more sites able to be modified than that of penicillin, people are more enthusiastic in semisynthetic cephalosporins than the work of semisynthetic penicillin. Semisynthetic cephalosporins target resistant strain and have increased activity, expanding the antimicrobial spectrum and reducing the side effects. Since 1960s, four generations of cephalosporins have been developed. Currently, more than 50 cephalosporins are used in clinic, such as cephalexin, cefazolin, cefotaxime, ceftazidime, ceftriaxone, cefpirome, cefepime, and cefprozil.

Today, β-lactam antibiotics (mainly penicillins and cephalosporins) have become the most important antibacterial infection drugs. In the pharmaceutical market, the sales of β-lactam antibiotics equal the sum of all other antibiotics.

Structural modification of tetracyclines and aminoglycosides

While the structural modification of β-lactam antibiotics was popular, the structural modification of tetracyclines and aminoglycosides was also conducted in full swing.

Chlortetracycline, oxytetracycline, and tetracycline and other tetracycline antibiotics have activities against Gram-positive and Gram-negative bacteria. However, with the emergence of resistant strains, the efficacy was significantly decreased, even led to treatment failure. In addition, tetracycline antibiotics have low water solubility, and this has made them difficult to produce suitable injection. Furthermore, their effective plasma concentration time is short. Thus, ammonia methylation of the amide group of tetracycline antibiotics can obtain aminomethyl tetracycline derivatives such as glycine tetracycline and morpholine tetracycline. These derivatives remedy the shortcomings, and have been used clinically. Chemical modifications were made to produce deoxy-oxytetracycline (doxycycline) and from demethyl-chlortetracycline to minocycline, which has the strongest antibacterial activities in the tetracycline class of antibiotics, and effective against resistant strains. Using oxytetracycline as raw materials to semisynthesize doxycycline can be 10 times more effective than tetracycline. Only one administration per day is required. Tigecycline was marketed in 2005. It is a structural modification derivative from minocycline and has strong activity against a variety of resistant strains.

Structural modification of aminoglycoside antibiotics is mainly targeted at drug-resistant mechanism. The idea is to eliminate the functional group that will be attached by inactivating the enzymes of drug-resistant bacteria, or to inhibit drug-resistant enzyme for effective derivatives screening. For example, 3'-hydroxyl of kanamycin is easy to be phosphorylated by resistant bacteria and lose antibacterial activity. After structural modification, 3'-deoxy-kanamycin is effective against resistant bacteria.

Structural modification of erythromycin

Erythromycin is a commonly used macrolides. Its main drawbacks are the instability in gastric acidic medium causing easily damaged and inactivated. This leads to low concentration in the blood. Again, it has strong bitterness. To overcome these drawbacks, chemical modifications have been performed for obtaining a series of derivatives that are more acid stable, and exhibit higher bioavailability and better adverse effects. After structure modification, 8-fluoro-erythromycin A is free from gastric damage and has higher plasma concentration, low liver toxicity. This is a good example for semisynthetic derivatives. Roxithromycin is another example. It was obtained from C9 carbonyl modification of erythrymycin and are much more stable as well as several fold higher plasma concentration. Currently, azithromycin, roxithromycin, clarithromycin, dirithromycin, and telithromycin are used in clinic as well. Especially, azithromycin achieved a great credit during 2003 "SARS" epidemic for excluding suspected patients. Only those fever patients do not work after taking azithromycin can be determined "suspected SARS patient" because azithromycin can control pneumonia caused by bacteria, mycoplasma, chlamydia, and rickettsia, which are typical pneumonia.

Since Prontosil (sulfanilamide) was founded by Domagk, it has been a chemotherapy era against bacterial infection. Then, penicillin was discovered by Fleming. This opens up an antibiotic era. Streptomycin discovered by Waksman brings the golden era of antibiotics. During 80 years, a large family of antibiotics has been available (Fig. 4-17). It is no exaggeration to say that the contribu-

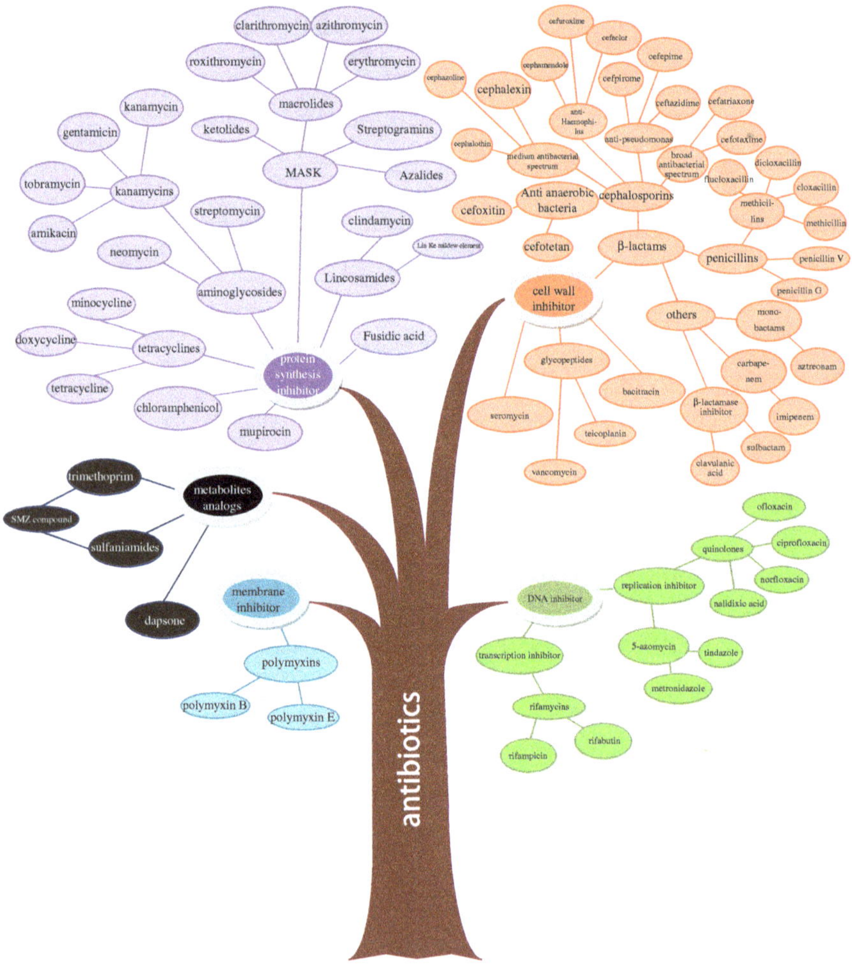

Figure 4-17: Antibiotics family.

tion it made to human life and health is incomparable to any other drugs. Today, each of us living on the planet should thank to these great scientists and to all contributing their efforts in antibiotics.

Medicines are toxic somehow

Adverse effects of antibacterial drugs

About 80 years ago, people died due to bacterial infection everywhere. The discovery and use of antibiotics brought the hope to the mankind and saved countless lives. However, everything has pros and cons and antibiotics are the same. They can also cause a lot of adverse reactions at the same time playing a therapeutic role.

Adverse effects of antibiotics are detrimental and nonrelated to the purpose of medication reactions caused by normal doses of antibiotics for the treatment of diseases. Adverse effects of antibiotics include toxic reactions, allergic reactions, and duo infections.

The main toxic effects of antibiotics are dose and time dependent. This usually refers to physiological biochemical abnormality and pathological changes caused by antibiotics, such as nervous system, kidneys, liver, blood system, gastrointestinal reactions, injection site local reactions, and bone development retardation. Antibiotics that cause damage to the nervous system are aminoglycoside antibiotics such as streptomycin, gentamicin, and kanamycin. These drugs can damage the auditory nerve and cause hearing loss, tinnitus, and vertigo. Ethambutol may damage optic nerve. Antibiotics that cause blood system damage such as chloramphenicol, sulfonamides, cephalosporins, and antitumor antibiotics. These drugs affect the regeneration of blood cells if used for a long-term large-scale administration. It usually reduces white blood cells, granulocytes, platelets and entire blood system, and damage the blood system. It can also cause aplastic anemia and hemolytic anemia. Antimicrobial drugs are usually absorbed by the intestine, metabolized in the liver, and excreted through the kidney. Therefore, liver and kidney are most vulnerable to be damaged by antibacterial drugs. Drugs that cause liver damage are tetracycline, rifampin, ketoconazole, and lincomycin. These drugs can cause jaundice, hepatocytes damage, liver damage, and even death in extreme situations. Drugs that cause kidney damage are kanamycin, cephalosporin, polymyxin B, and sulfaniamide drugs.

Allergies are the most common reactions to the use of antimicrobial drugs. Common allergic symptoms are fever, rash, joint pain, bronchial spasms, asthma, purpura, and bleeding, an anaphylactic shock that can occur in severe cases. Some antimicrobials need skin test before administration. Its purpose is to prevent potential anaphylactic shock. For those are easy to cause severe allergic reaction need to pay special attention when using the drugs. Once an allergic reaction was observed, drugs should be discontinued immediately.

Duo infection is caused by large scale and long-term use of antibiotics such as tetracycline, cephalosporin, and chloramphenicol. Then, the body's normal flora balance was destroyed. The imbalanced gut flora can cause new infections in the course of antimicrobial agents.

It is clear that antimicrobial drugs can cure the disease, but improper use can cause the disease, or even cause threat to life. Therefore, antimicrobial drugs must be used under the guidance of a doctor. Do not use excessively and indiscriminately.

Penicillin and anaphylactic shock

Penicillin is most likely to cause allergic reactions. About 5–6% of humans are allergic to penicillin, and the chance of anaphylactic shock is also the highest. Allergic reactions can occur at any age, at any dose, or any way you use it. The allergy occurs fast, as an allergic rash, urticaria, drug rash, swollen lymph nodes in nonsevere case, and anaphylactic shock in severe scenario. Dyspnea, cyanosis, cold sweats, cold extremities, blood pressure fall, convulsions, and coma are the common symptoms. If no action is made promptly, the effect will be life threatening.

Some people are allergic to direct or indirect contact with very small amounts of penicillin, such as drinking milk from penicillin injected cows, the use of penicillin contaminated syringes, working in penicillin-found environment, or suffering from skin mycosis (produces substances similar to penicillin) and cause sensitization. There has been a report that a person allergic to penicillin passed around a patient receiving penicillin injection made him shock. In 1950s, China has the event that the injection of streptomycin has led to the death of more than 50 people. The final investigation revealed that the plant, which produces streptomycin was originally used for penicillin production. Governments of different countries now have clear regulations that the plant for penicillin production must be independent in order to prevent contamination of penicillin in other drugs.

Streptomycin and nervous system damage

Streptomycin can cause nervous system damage to some people. As the symptoms include skin rash, urticaria, erythema, measles-like rash, scarlet fever rash, pemphigus-like rash, eczema-like rash, purpura and vascular edema, or anaphylactic shock in severe case.

Streptomycin is very toxic to kidneys, but very few people are allergic to streptomycin. Fever, drug rash in nonsevere case, exfoliative dermatitis (a serious systemic skin disease), or even anaphylactic shock in severe case are the common symptoms.

The most serious danger of streptomycin is the damage to nervous system. Streptomycin is easy to damage the eighth cranial nerve (auditory nerve), cause dizziness, ataxia, tinnitus, hearing loss, and permanent deafness in severe case. Many deaf people can trace back to the use of these drugs during childhood or juvenile-year period.

Erythromycin and occasional gastrointestinal irritation

Erythromycin can cause allergy occasionally. Systematic allergic itching, skin rash, pimples, or herpes in severe case are the symptoms. The most common adverse reactions are gastrointestinal reactions such as nausea, vomiting, abdominal pain, and diarrhea. Using drugs after meal can alleviate gastrointestinal reactions.

Erythromycin propionate that can cause gallbladder, bile duct, and hepatic bile duct cholestasis leads to induced liver disease and jaundice, direct liver toxicity, and liver cell inflammation as well. Therefore, erythromycin is contraindicated in pediatric hepatobiliary or hypoplasia, which caused kidney excretion dysfunction and pregnant women, lactating women, patients with liver damage.

Since erythromycin is insoluble in water, injections are often used in glucose solution to prevent precipitation. However, erythromycin in glucose solution can lead to vascular irritation. Prolonged vascular access leads to intravenous damaged causing intimal cells to fall off, and the fibers in the vessel wall to harden. Therefore, erythromycin cannot be given to patients with phlebitis, and prolonged injection of erythromycin into the same vein should be avoided.

Tetracycline and "tetracycline teeth"

Tetracycline antibiotics generally have low toxicity in clinical use. Sometimes they cause nausea, vomiting, abdominal discomfort, gastrointestinal inflatable, diarrhea and other intestinal tract reactions, and drug fever and rash occasionally.

The biggest problem is "tetracycline teeth". Taking tetracycline during the mineralization of tooth development, tetracycline and calcium molecules inside the tooth structure can form very stable chelates deposited on the tooth structure. This produces colors on the teeth and cause incomplete enamel development. Initially, the color is yellow, showing yellow fluorescence in the sunlight. The color will gradually become deeper and keeps the yellow over a longer period of time. China had widespread use of tetracycline antibiotics in 1970s and 1980s. Because of this, many children had "tetracycline teeth" in that era. Moreover, the colors were very difficult to remove in the future. Although there is no excessive damage to the body, it can also be regarded as a branded print during the antibiotic era.

Quinolones and cartilage developmental disorders

Quinolones can cause allergic reactions, and damage to gastrointestinal system, kidneys, liver, and central nervous system, but at lower occasions.

Serious adverse reactions of these drugs are related to cartilage development. They are easily absorbed by the cartilage tissue; then, they deposit in the bone marrow. This could damage the development of the cartilage cells, and affect the skeletal development of children and fetus. Therefore, pregnant women and children under 12 are prohibited to use this kind of drugs. Breastfeeding should be stopped if this drug should be used during the lactation period. Long-term use for adult can lead to joint diseases such as increasing joint mass, stiffness, and limitation of motion.

Few quinolones (e.g., lomefloxacin) have a significant photosensitivity feature. After exposure to the light, the drug can produce some toxic substances due to chemical reactions. Taking these drugs may induce toxic substances even without direct sunlight, but they may be more severe under the sunlight. Therefore, pharmacists should remind the patients to avoid the sun light after taking the drug, and do not accept the artificial UV exposure.

An extreme example—thalidomide

Thalidomide was once notorious in the pharmaceutical world! Because, it is the cause of "seal fetus incident." Meanwhile, the seal fetus incident also made the world's most famous, the most important article—"Kefauver Harris Amendment" to be legislated. It provides that US Food and Drug Administration (FDA) should not only need to demonstrate its safety and efficacy before the launch of the new drug, but also need to pass the same rigorous scientific testing for the drugs already on the market.

In 1953, Ciba Pharmaceuticals, the predecessor of Novartis Switzerland, first synthesized thalidomide. They originally intended to develop new antimicrobial drugs, but pharmacological tests showed that thalidomide has no antibacterial activity. Ciba gave up the R&D of thalidomide, which later studied by Chemie Grünenthal pharmaceutical company, Federal Republic of Germany after Ciba abandoned thalidomide. They put great efforts and human resources into the study of thalidomide on the central nervous system and found that the compound has a certain sedative and hypnotic activity, but more significantly, can inhibit the reaction of pregnancy (antiemetic activity and others). In October 1957, thalidomide (as the reaction of pregnancy stopped after taking thalidomide) officially markets in the Europe. In less than 1 year, thalidomide is popular in Europe, Africa, Australia, and Latin America. As an "anti-pregnancy reaction drug with no adverse effects," it becomes the "ideal choice for pregnant women."

However, when thalidomide entered the United States, it got trouble. A small US pharmaceutical company Merrill was licensed to distribute "thalidomide" in the US. In 1960, they applied to FDA for market approval. At that time, FDA officer, Francis Kelsey, was responsible for the application. She noticed that thalidomide can cause hypnotic effect in human but not in animal studies. Did this mean that human and animals have different pharmacological response to this drug? Almost all of the drug's safety evaluation results came from the animal study. Are they reliable? Kelsey noticed that there are medical reports showing that the drug has caused paralysis and some patients taking the drug feel tingling fingers. She then speculated that the drug will have adverse reactions in pregnant women and affect fetal development. Merrill replied that they had studied the drug on pregnant rats and pregnant women, but no adverse effects. But, Kelsey insisted on more research data, which caused Merrill's dissatisfaction. She suffered a lot of accusations and unwarranted pressure.

When the two sides were wrangling, Australian obstetrician William McBride reported in the British "Lancet" claimed that "thalidomide" can cause birth defects. Mothers who delivered in McBride's clinic gave birth of babies suffering from a rare malformation symptom, phocomelia, with limb hypoplasia, short just like the seals pinnipeds. These women have all taken "thalidomide." In fact, at this time, more than 8000 phocomelia babies have been found in Europe and Canada. McBride is the first to link with "thalidomide." Toxicology studies later showed that thalidomide shows a strong teratogenicity to primate. Different from human, rats lack an enzyme *in vivo* to transform "thalidomide" into a harmful isotype. For this reason, no teratogenicity was observed in rats. The adverse reactions of thalidomide occurs in early pregnancy (first 3 months of pregnancy), a period when the baby develops limbs. From November 1961, "thalidomide" has been forced to withdraw in the world. Merrill companies also withdrew their application. After a long legal battle, the German company Chemie Grünenthal responsible for research and development of thalidomide paid 110 million DM for compensation, and finally went to bankruptcy.

The thalidomide incidence is a tragedy in the pharmaceutical history. At least 8000 Infant deformities were caused by taking thalidomide; another 5000–7000 babies died before birth due to deformities. The United States was not affected in particular because the FDA officer Dr. Kelsey insisted on the review of thalidomide's application. President Kennedy awarded Kelsey the "Outstanding Federal Citizen Service Medal." Because of thalidomide incidence, the public asked Congress to strengthen the legislation. On October 10, 1962, Congress passed the "Kefauver Harris Amendment." FDA thus became the world's most authoritative food and drug inspection agency.

Later, thalidomide was found rapid anti-inflammatory effects and efficacy for leprosy patients with nodular erythema. This conclusion was proven by 90% patients with erythema nodosum leprosy. In 1998, thalidomide was recommended for ENL patients by the US FDA. In 2006, FDA reviewed and approved thalidomide to treat multiple myeloma or myeloma.

Thalidomide was also recognized in China by the Chinese Medical Association: In addition to treat erythema nodosum leprosy, thalidomide can treat multiple myeloma or myeloma according to the "Clinical Practice Guidelines, Hematology" and treat ankylosing spondylitis and Behcet's disease according to the "clinical Practice guidelines, rheumatology."

In addition, in recent years, thalidomide has achieved gratifying and encouraging results in the treatment of immune, inflammatory, inhibiting angiogenesis and some other difficult and complicated diseases. It is the renascence of new thalidomide.

Chapter Five

A Protracted Tug of War

"Strategy and Tactics" for Drugs to Kill Bacteria

Precise knowledge of self and the threat leads to victory

Perhaps you've visited a considerable number of ancient cities, and have observed all types of wonders. But, have you ever been to the bacterial "castle"? After hundreds of millions of years of evolution, in this sophisticated castle of life, there are no rumbling motors or spinning wheels, but only quiet, ordered, and efficient operation.

Precise knowledge of self and the threat leads to victory. How is the bacteria castle constructed? How does the bacteria castle operate?

In Chap. 1, we have mentioned that bacteria appear like eggs. This has already defined the basic structure of the bacteria. Here, bacteria are also compared to a castle (Fig. 5-1). We use this castle to describe its function and to defend the antibiotic attack. The bacterial cell wall can fix the outer shape of the cell. Moreover, it protects the cell, which is like the solid city walls; close to the inside of the cell wall is the cell membrane, which controls the absorption of the nutrients and secretion of the metabolites. It resembles the castle choke; the major genetic component DNA is located in the central core area. It is like the castle headquarters to give command and take decision that are effectively implemented by a cascade of chain pathway. This makes all the works properly functioned so that bacteria are under normal growth and reproduction.

In the operation of bacteria castle, proteins are important materials for life and metabolic activities. The synthesis of protein is

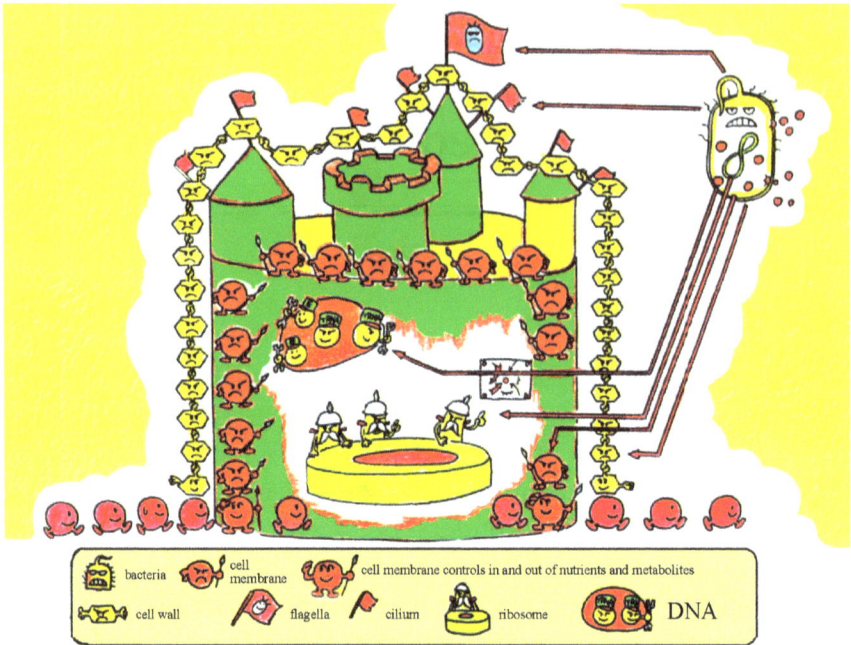

Figure 5-1: Bacteria castle.

controlled by genetic information from DNA. Every three nucleo-tides in DNA chains determine an amino acid. Amino acids join together to form a certain protein. Moreover, protein appears like a train, where each amino acid resembles a coach; this is the "primary structure" of a protein. This train is functional only after the correct folding and is able to play the biological activities. This is the "higher-order structure" of the protein after folding. Different fold-ing structures are referred to as "secondary structure," "tertiary structure," and "quaternary structure." Once protein synthesis is blocked, the bacteria cannot perform normal life activities. It loses the ability to survive and reproduce. Therefore, how does the head-quarters give commands to guide protein synthesis?

The first step is to transcribe the information on the DNA into a code. This will open up the double strands of the DNA. Then, RNA is synthesized by using one strand as the template, which is called transcription. RNA after transcription has three subsets. RNA with all the information for protein synthesis is called messenger RNA (mRNA) because it acts like a messenger. The second one is ribosomal RNA (rRNA). rRNA and certain proteins can assemble into a small particle called "ribosomes." Proteins are synthesized in these small particles. For this reason, the ribosomes are said to be the "factory" of protein synthesis in the cells. Moreover, transfer RNA (tRNA) is required to translate mRNA into a protein. rRNA should first recognize not only the information on the mRNA — the genetic code, but also the text of protein — amino acids, to perform this "translation." Second, tRNA as a transporter can follow the commands from the headquarters to find certain particular amino acids and lead to ribosome to match with mRNA. There are many "craftsmen" — enzymes in protein synthesis as well. The genetic information in DNA is passed on follow this rule: DNA is first transcribed into mRNA, with the participation of rRNA and tRNA, this information is then translated into protein. This is the "central dogma" in genetics. Figure 5-2 depicts how bacteria headquarters give commands to guide the protein synthesis.

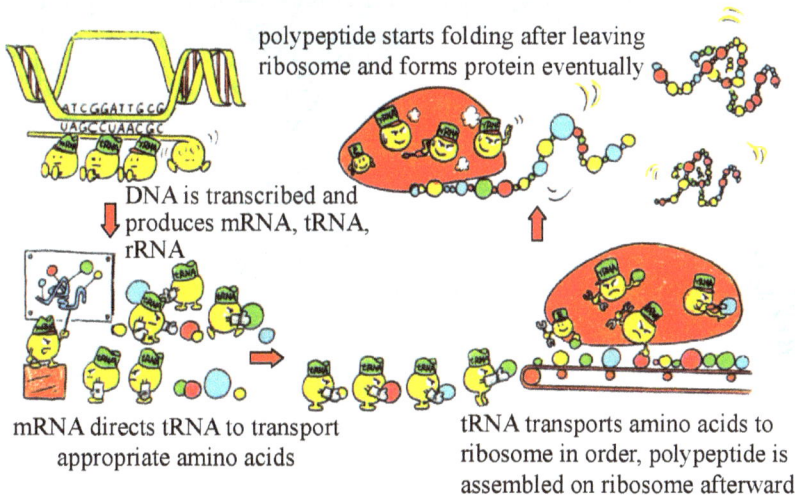

polypeptide starts folding after leaving ribosome and forms protein eventually

DNA is transcribed and produces mRNA, tRNA, rRNA

mRNA directs tRNA to transport appropriate amino acids

tRNA transports amino acids to ribosome in order, polypeptide is assembled on ribosome afterward

Figure 5-2: Transmitting commands from "bacteria headquarters" and controlling protein synthesis.

Most fortified castle is not unassailable. If the fortifications of the bacteria castle (cell wall, cell membrane) were damaged, if the commands issued by the headquarters of the bacteria (DNA replication, RNA transcription and protein translation) is corrupted or is hard hit then the castle will be collapsed and the bacteria will be destroyed. Since Fleming's discovery of penicillin, scientists have found different kinds of antibiotics. They target on different mechanisms of the bacteria to inhibit their growth and kill the bacteria eventually.

Destroy the castle wall bricks of the bacteria

Cell wall is an important component of the bacteria castle. It is constructed by numerous craftsmen (enzyme) to bring a variety of bricks (most important of which is the peptidoglycan) together.

One of the craftsmen, "penicillin-binding protein (PBP)", is the key to build the walls of the bacteria castle. If we can target this key protein and deactivate its function, then building the wall will be hindered causing defects to the walls. The internal contents of bacteria have various substances and high osmotic pressure. Water seep inside to make the cell swell like a balloon. If there are pores in the cell wall surface, the internal contents will gush out leaving only a shell of gas discharged. This is called lysis, which causes bacterial cell death. Penicillin antibiotics can kill bacteria by inhibiting their cell wall synthesis and bacterial growth (Fig. 5-3).

For this battle tactic, we should note the following: 1. The timing of the attack must be appropriate. Penicillin is more active against bacteria under log phase growth state than bacteria under stationary growth state because the former requires constant synthesis of new cell wall while the already-synthesized cell wall of the latter is unaffected. 2. The structure of the cell wall has to be simple. If trenches are present outside the walls, such as the outer membrane layer of Gram-negative bacteria cell wall, it is more difficult to capture the key protein and reduce the killing effects.

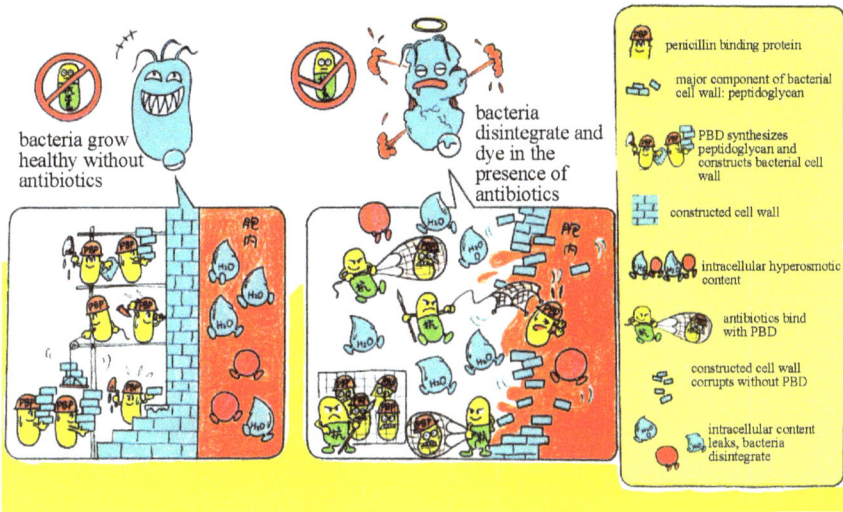

Figure 5-3: Antibacterial mechanism of penicillin.

Strangling bacteria's "choke points"

Cell membrane can be described as the bacterial castle choke. It protects a variety of elite troops — proteins and enzymes. Some can protect the cells by maintaining the normal intracellular osmotic pressure; some are responsible for controlling the absorption of nutrients outside and transporting metabolic products; some produce various raw materials for cell wall synthesis; some provide energy for cellular activities.

There are several tactics to attack membrane.

Antimicrobial drugs such as daptomycin, polymyxin, and tyrothricin can punch a hole in the bacterial cell membrane to form an ion channel so that the internal contents will leak to the extracellular environment. The leaking content is related to the type of antibiotic, its concentration, and administrating time, for example potassium, inorganic phosphorus, organic phosphorus, amino acids, nucleic acids, proteins, which have a fatal effect on the cells.

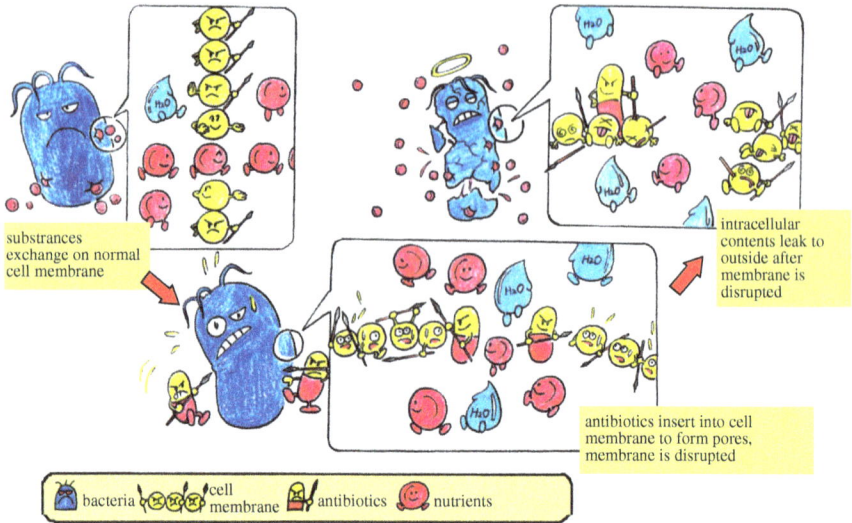

Figure 5-4: Intracellular contents leak after cell membrane was disrupted by antibiotics.

The components of bacterial cell membrane are predominantly lipid soluble. If the striking force — antimicrobial agents have similar lipid-soluble characteristics with cell membrane, they can easily pass through the phospholipid bilayer of the cell membrane, like wearing the camouflage clothing, and directly reach the target site to deliver the efficacy.

Antimicrobial drug can interact with cell membranes or cover the membrane like a "carpet" to destroy the function of the cell membrane. Alternatively, it can punch a hole on the membrane causing leak of internal contents to kill the bacteria (Fig. 5-4).

Occupying bacteria "headquarters"

DNA is the fundamental base of cell structure and life activities in bacteria cells. It is considered as the headquarters of the bacteria castle. DNA is made of different combinations and arrangements of four different nucleotides in two long-chain spiral, known as a double helix. All the genetic information is passed on to the next generation by DNA replication while bacteria go under reproduction (Fig. 5-5).

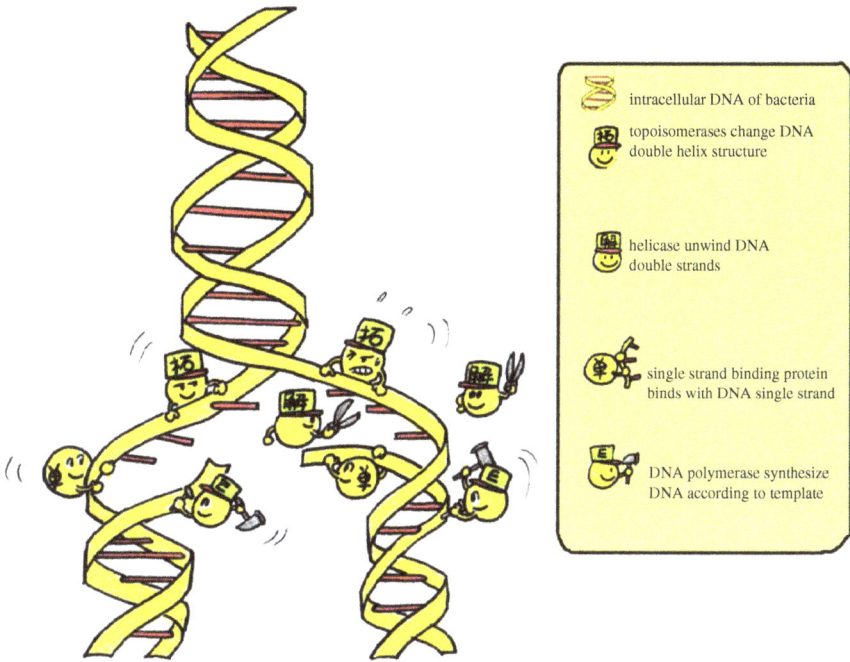

Figure 5-5: DNA replication.

DNA replication starts from gradually unraveling the double helix. A complement chain is synthesized according to each parental strand as a template. This is like reproducing a staircase. Two handrails are taken apart as the template; then, another handrail is produced identically according to the template; finally, two pairs of helical staircases are made. DNA replication strictly follows this procedure to ensure that the genetic information of parents can accurately transfer to offspring.

A large number of "artisans" (i.e., various enzymes and related protein cytokines) are involved in DNA replication. The process can be divided into four stages. The first stage is to make changes to the normal spiral structure, as well as unraveling double strands. The so-called DNA topoisomerase is responsible for changing the DNA double helix (or supercoiled) spatial structure in order to facilitate unraveling double strands. DNA helicase will unlock the double strands to be used as templates for new DNA strand replication. Then, the single-strand binding protein (also known as DNA-binding protein) closely binds the single-strand DNA for stabilizing the single-strand template in order to facilitate its role as a template. The single-strand binding protein can also bind to the new single-strand DNA to protect them from hydrolysis by the "bad guy" nucleases. The second stage is initiation of replication: DNA replication is triggered by a complicated stage. Several craftsmen including primase and initiator precursors participate in this stage. The third stage is the extension of the DNA chain. This stage is charged by DNA polymerase, which can correctly synthesize the DNA according to the template. The fourth stage is termination. Since there is a termination region in a particular location of the DNA templates, all the skilled craftsmen mentioned above can do nothing at this stage that leads to the termination of replication. After DNA replication, the structure of the DNA can be changed to helix by topoisomerase and proceed to further assembly.

Quinolone antibiotics, such as nalidixic acid, pyrrole acid, and Piperazine acid, can inhibit DNA replication by inhibiting the bacterial DNA gyrase or topoisomerase causing death of bacteria (Fig. 5-6).

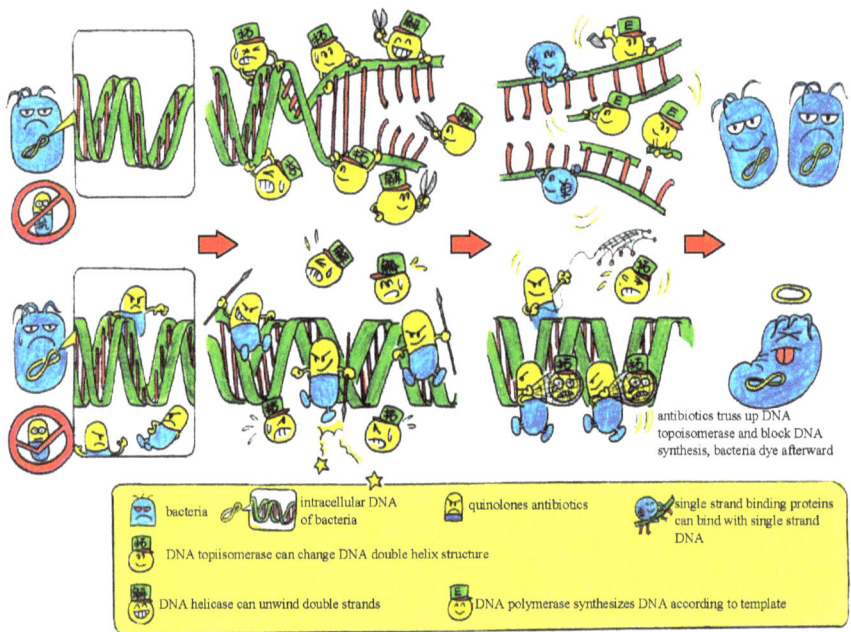

Figure 5-6: Anti-helicase and anti-topoisomerase mechanism of antibiotics.

The antibacterial mechanism of metronidazole is also inhibiting DNA replication.

Destruction of "secret codes"

Double-stranded DNA is opened up by helicase so that all the codes can be observed by "transporter." Then, the codes were passed to the messenger RNA by transcription. The follow-up, protein synthesis, is guided by mRNA and the work is performed by RNA polymerase, like "a courier."

Rifamycin antibiotics use this mechanism to closely bind with the RNA polymerase to prevent the transfer of DNA secret codes. Several binding sites existed in the structure of rifamycin antibiotics, exhibiting strong activities. For example, rifamycin antibiotics can interact with *Mycobacterium tuberculosis* RNA polymerase and create magnetic forces to form a stable complex and inhibit the activity of the enzyme, block the initiation of RNA synthesis, and lead to the inhibition of RNA synthesis. Figure 5-7 shows the destruction of secret codes during DNA transcription by rifamycin antibiotics.

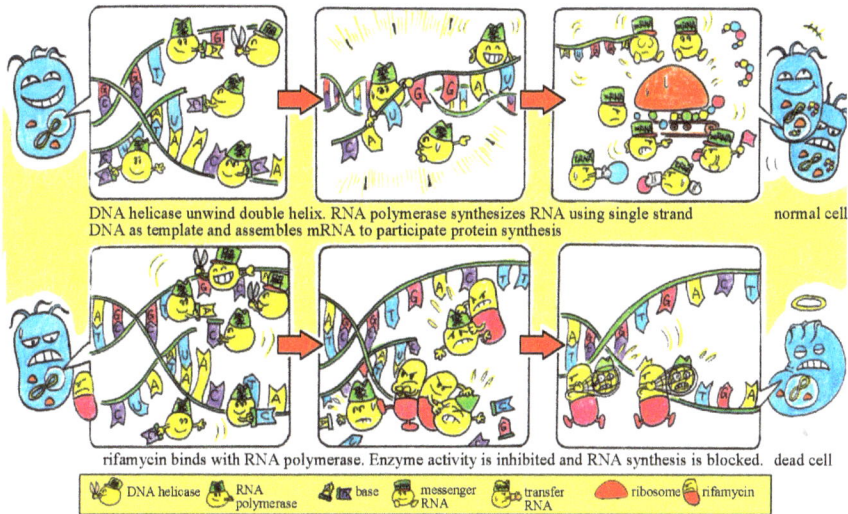

DNA helicase unwind double helix. RNA polymerase synthesizes RNA using single strand DNA as template and assembles mRNA to participate protein synthesis — normal cell

rifamycin binds with RNA polymerase. Enzyme activity is inhibited and RNA synthesis is blocked. dead cell

DNA helicase · RNA polymerase · base · messenger RNA · transfer RNA · ribosome · rifamycin

Figure 5-7: Rifamycins destroy copying of DNA codes.

Ambushing protein "Factory"

Protein synthesis is occurred in ribosome where every amino acid is linked one by one followed by a series of modification to manufacture functional proteins. Thus, ribosome is considered to be the factory of bacteria proteins.

To defeat the enemies, not only the power, but also smart is needed. Many clinical used antibiotics can inhibit the growth of bacteria through ambushing protein factory, such as penicillin, gentamicin, chloramphenicol, erythromycin, tetracycline, lincomycin, and sulfanilamide. However, for the ambushing protein factory, different antibiotics use different tactics to achieve their goals. Some typical strategies are described as follows.

Siege the protein synthesis

One set of codons determining protein synthesis existed in mRNA (messenger RNA), receiving secret codes. With the help of tRNA (transfer RNA), interpreter, and transporter, protein is translated in ribosomes, the factory of protein synthesis. This process is known as protein translation.

Erythromycin and tetracycline antibiotics are considerably effective protein synthesis inhibitors. They target the key enzyme, peptidyl transferase, on ribosome responsible for capturing and transporting amino acids. Consequently, amino acids cannot attach to the locomotive like normal coaches. Since protein synthesis is blocked, essential life activities are interrupted and the growth of bacteria is inhibited.

After streptomycin entering into the bacterial cells, they can disrupt the proofreading process in translation causing mismatch codons. In this manner, streptomycin can block the correct initiation of protein synthesis, or interfere with the elongation of the newly synthesized strand to create abnormal proteins. When abnormal proteins integrate, it will cause changes to the permeability and promote transportation levels for more aminoglycosides. With increase in the amount of antibiotics in the cell, the destruction of protein synthesis is exacerbated, resulting in the disintegration of the membrane structure and permeability damage, ultimately leading to cell death (Fig. 5-8).

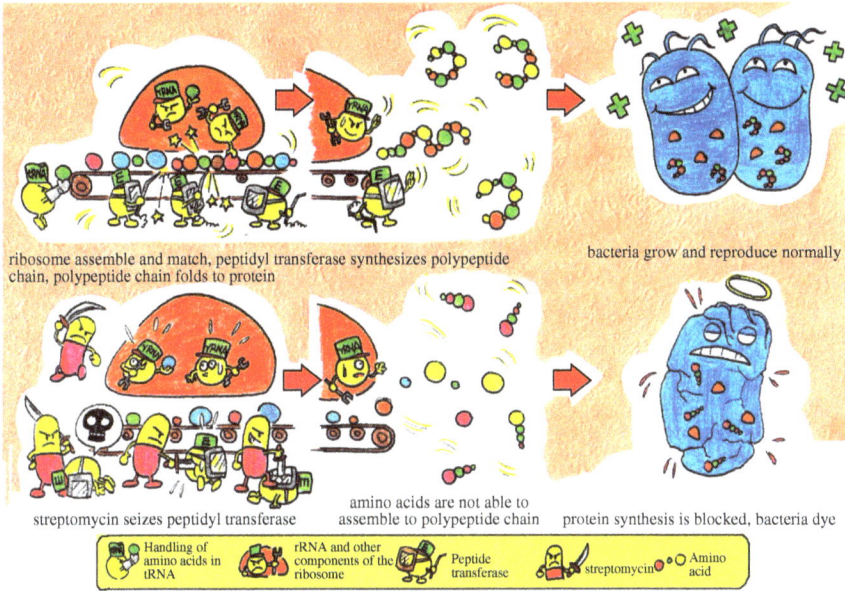

ribosome assemble and match, peptidyl transferase synthesizes polypeptide chain, polypeptide chain folds to protein

bacteria grow and reproduce normally

streptomycin seizes peptidyl transferase

amino acids are not able to assemble to polypeptide chain

protein synthesis is blocked, bacteria dye

| Handling of amino acids in tRNA | rRNA and other components of the ribosome | Peptide transferase | streptomycin | Amino acid |

Figure 5-8: Streptomycins attack protein synthesis.

Destruction of the protein factory

Complete factory for bacteria protein need assembly of a 30S ribosome subunit and a 50S ribosome subunit to become a 70S large ribosome. This is a regular protein factory for construction and manufacturing of proteins. New antimicrobial drug, Linezolids, can inhibit protein synthesis by inhibiting the formation of 70S large ribosome. Figure 5-9 shows that 50S ribosome subunit and 30S ribosome subunit cannot assemble into a regular protein factory in the presence of antibiotics linezolid.

Intercept "resource transporters" of bacteria

Folic acid is an important substance necessary for all life processes. The human beings acquire folic acid necessary for life from food. However, bacteria require de novo synthesis of the necessary folic acid. Folic acid is the resource for coenzyme F synthesis in creatures. Coenzyme F is the raw material for the synthesis of purine or

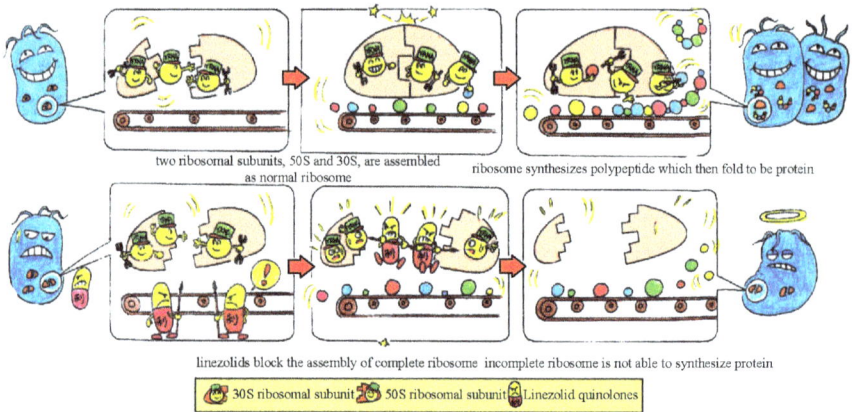

two ribosomal subunits, 50S and 30S, are assembled as normal ribosome

ribosome synthesizes polypeptide which then fold to be protein

linezolids block the assembly of complete ribosome incomplete ribosome is not able to synthesize protein

30S ribosomal subunit   50S ribosomal subunit   Linezolid quinolones

Figure 5-9: Antibacterial linezolid destroys the construction of protein factory.

pyrimidine bases that are necessary for DNA synthesis. During the synthesis of folic acid, there are several "transporters," folic acid synthase, that are involved in the process. In normal situation, these transporters can obtain required resources for folic acid synthesis from the cell.

Scientists found that sulfanilamide can pretend itself considerably similar with para-aminobenzoic acid, the resource for folic acid synthesis. Thus, in the presence of sulfanilamide, those transporters in the cells will transport sulfanilamide instead of the real resources, para-aminobenzoic acid. It appears as if the transporters are being blocked, and the synthesis of folic acid is stopped. For this reason, bacteria can no longer survive and reproduce. As a matter of fact, all enzymes involved in biosynthesis in cells are smart transporters. They are not simple labors, but are "craftsmen." They are able to correctly transport the desired "resources" to the designated place (Fig. 5-10).

In short, although the mechanisms of action of different antibiotics against bacteria are not the same, they all disturb or hinder their growth and reproduction to kill or inhibit the growth of bacteria.

Bacteria's Fight Back Against the Drugs — Resistance

God created everyone differently. The development of nature follows a principle that never has been changed. "There must be a

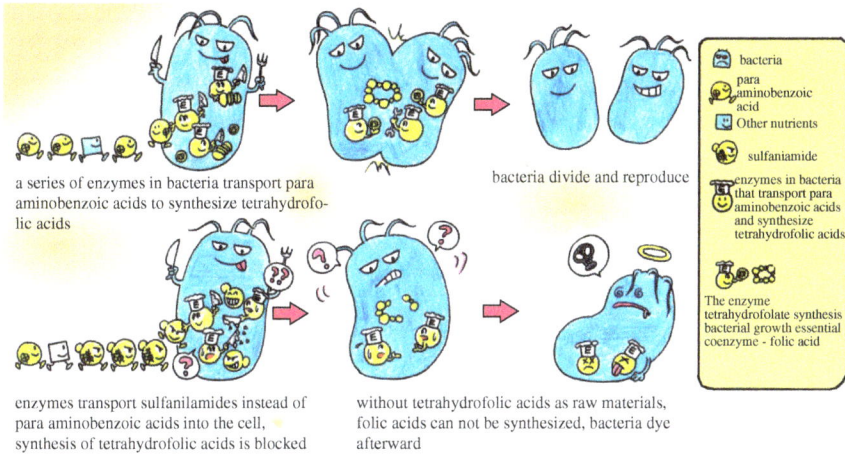

a series of enzymes in bacteria transport para aminobenzoic acids to synthesize tetrahydrofolic acids

bacteria divide and reproduce

enzymes transport sulfanilamides instead of para aminobenzoic acids into the cell, synthesis of tetrahydrofolic acids is blocked

without tetrahydrofolic acids as raw materials, folic acids can not be synthesized, bacteria dye afterward

bacteria

para aminobenzoic acid

Other nutrients

sulfaniamide

enzymes in bacteria that transport para aminobenzoic acids and synthesize tetrahydrofolic acids

The enzyme tetrahydrofolate synthesis bacterial growth essential coenzyme - folic acid

Figure 5-10: Sulfanilamides intercept resource transporters of bacteria.

shield along with a spear." This principle has been proved in the history of human civilization for thousands of years, particularly the history of fight. Similarly, this is also true in the endless war between human and bacteria. We have tried every effort to develop newer and more effective antibacterial drugs — antibiotics; meanwhile, bacteria have constantly evolved their "body", fully inclusive and equitable, to create a better updated "shield" to resist the human invention of new weapons. The sharper the spear, the stronger the shield. In this desperate war, when human feel that their spear is sharper when new antibiotics are invented, bacteria has already started making stronger shield to protect themselves. Bacteria resistance has been an unprecedented threat to human life and health. Those resistant bacteria that are hard to kill in clinic are known as "superbugs."

The emerging and development of resistant bacteria

Although the application of sulfanilamide is earlier than penicillin, the introduction of penicillin and other antibiotics had made the clinical bacterial infection to be well controlled and the average human life expectancy to be extended to 15–20 years. However, the clinical application of antimicrobial drugs is accompanied by the development of bacterial resistance. Moreover, the emergence of bacterial resistance has become faster and faster, the degree of resistance has become more and more severe, the type of resistant strains has become more and more, and the frequency of drug resistance has become more and more often. Data shows that in 1950s and 1960s, bacteria that can be controlled by 20–30 thousand units of drug now requires hundreds of thousands or even millions of units to treat. Staphylococcus, enterobacteria, tuberculosis, and Shigella have been the life-threatening infections of human for such a long time because they have stronger and stronger "shield" continuously.

In 1940, when penicillin was just introduced into clinic, it was able to kill all of *Staphylococcus aureus*. In 1942, penicillin-resistant *S. aureus* emerged. In 1944, scientists found penicillinase (can destroy penicillin, also known as β-lactamases) in *Staphylococcus aureus* is responsible for the resistance. In 1947, the first case of clinical drug-resistant bacteria was found. Because penicillin showed magic effects in the first few years, abuse use of penicillin that occurred in the hospitals resulted in the spread of resistant bacteria in the hospital in the mid-1950s. About 90% of Staphylococcus is resistant to penicillin.

In order to solve bacterial resistance to penicillin, scientists began to study the development of semi-synthetic penicillin that is resistant to penicillinase. In 1960, methicillin, a semi-synthetic penicillin, that cannot be hydrolyzed by penicillinase was discovered and used in clinic. But, soon later methicillin-resistant *Staphylococcus aureus* (MRSA) emerged.

Glycopeptide antibiotic vancomycin is frequently used in clinic when β-lactam antibiotics or other antimicrobial drugs fail. It is considered to be the last line of defense and the "elite Antibiotics" in the treatment of antibacterial infection. However, in July 2002, the world's first case of vancomycin-resistant *Staphylococcus aureus* (VRSA) was found in the wound of a patient from Michigan, US. We still have no effective drugs to clinically treat this "superbug."

1960s is the peak time for the development of antibiotics. Afterwards, the discovery of new antibiotics has slowed down. But, the type and drug-resistant mechanisms of bacteria are still changing. Currently, several bacteria are resistant to several antibiotics (multi-drug-resistant bacteria). Because of the excessive use of antibiotics, the resistant bacteria spread rapidly and have become a global issue. Currently, about 70% of the bacterial infection is resistant to at least one commonly used antimicrobial drug. Some bacteria are not sensitive to all antibiotics and can only rely on surgery or toxic drugs.

Bacteria have evolved during the fight against antimicrobial drugs. The way they can be evaded from or resist directly against drugs is bacterial resistance! Our dependence and abuse use of antibiotics make the situation even worse. They selectively retain resistant population, and make that population become majority. Bacterial resistance follows the survival of the fittest law of the nature.

Abuse use of antibiotics is responsible for bacterial resistance

Everything in the universe is in transition. Change is eternal, not changing is temporal. Darwin's theory of evolution illustrates from macro-environment that all creatures are under slow evolution and the survivors are the fittest. But, it did not explain the essence that dominates the evolution. Currently, scientific findings are able to clearly answer the question — the essence that dominates the evolution is the genetic material DNA, that is, the genes located on the DNA to determine biological characteristics. Gene mutation changes the nature of the organism.

Natural evolution is considerably slow. It is estimated that, under natural conditions, only one bacterium of 1–10 million bacteria may become resistant owing to gene mutation. In general, it is similar to one drop of water in the sea and one piece of rice in the barn. It is difficult to get into trouble with such a small amount. But, when antibiotics were discovered and found to have antibacterial activities, conscious or unconscious abuse use of antibiotics has changed the situation. Tragedy of human has begun: a large number of sensitive bacteria are killed, the normal flora in human body is destroyed, and these resistant bacteria started to multiply rapidly. Particularly, when people found that a considerable number of antibiotics can promote the growth of animals and recognized this as a "scientific achievement", widely used as feed additives in short times to make huge economic benefits, the tragedy has begun to intensify. Moreover, the abuse use of antibiotics on the crop is another "culprit", causing resistant bacteria.

Why the abuse use of antibiotics leads to bacterial resistance? The following scientific discoveries can be strong evidences.

First: Long-term large amount of clinical use of antibiotics has an enormous selection pressure on bacteria. Most of the sensitive bacteria are killed so that those resistant bacteria who had only a considerably small proportion (10^{-9} to 10^{-6} spontaneous mutation frequency) multiply rapidly.

This is similar to the case without the use of antibiotics, only one out of one to a hundred million bacteria spontaneously changed from sensitive to resistant. The resistant bacteria are considerably few to stir the market. They are concealed in the huge amount of sensitive bacteria. Thus, a handful of these "enemies" of our body does not cause fatal injuries. When a large amount of sensitive bacteria was killed, this small group of "enemies" took the opportunity to propagate massively. Ultimately, when we used the same antibiotic, we found that it lost its activity (Fig. 5-11). Therefore, the less the amount of antibiotic the better is the activity. Moreover, the less the time taken to use the antibiotics, the better is the activity. Otherwise, it will increase the likelihood of a large number of resistant bacteria.

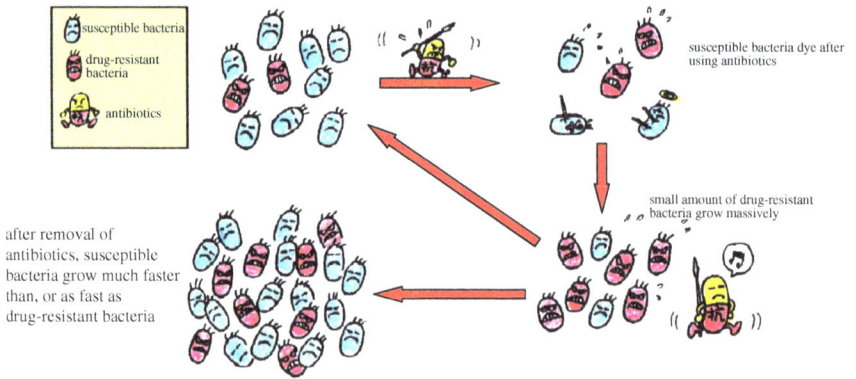

Figure 5-11: Resistant bacteria propagate massively under the pressure of antibiotics.

Extremely small amount of resistant bacteria are covered under a large number of sensitive bacteria. In the presence of antibiotics, sensitive bacteria are killed and extremely small amounts of resistant bacteria take the opportunity to propagate massively. After the usage of antibiotics is stopped, sensitive bacteria started growing again, and its propagation speed may be similar with or even way faster than the resistant bacteria.

In addition, nonclinical long-term use of antibiotics also has a huge selection pressure on bacteria. Particularly, antibiotics are widely used as "animal growth promoters" recently. This causes cross-resistance of bacteria in humans and animals. At the same time, a large number of antibiotics are used for crops. Thus, it increases the chances of resistant bacteria spreading between foods and humans. Animals are likely to be a reservoir of resistant bacteria and transfer to humans. On November 17, 2006, Shanghai government tested turbot in the market and found drug residues exceeding the limit. Aquatic product market issued a notice to stop selling turbot immediately because "all the turbots in the market have all been tested to contain nitrofuran metabolites, ciprofloxacin, malachite green, oxytetracycline, and other drugs, at varying degrees." Furthermore, wastes containing a small amount of residual antibiotic by manufacturer during production, as well as residues of antibiotics from hospital being released into the environment after clinical use, are also major causes of the spread of resistant bacteria and transmission.

Figure 5-12 shows how the residues of antibiotic enter animals, plants, humans, and environment and cause the development of bacterial resistance under selected pressure. Therefore, nowadays, while we enjoy the delicious food on the table, on the other hand, we have ingested a large number of residues of antibiotics into our body together with the delicious food. Many European countries and the United States have begun to prohibit the usage of antibiotics as growth promoters. Although there are many related provisions in China, illegal infringements occur occasionally in farms and antibiotics manufacturers. It is necessary to bring these to our attention.

Second: Antibiotics can induce bacterial resistance from the following aspects:

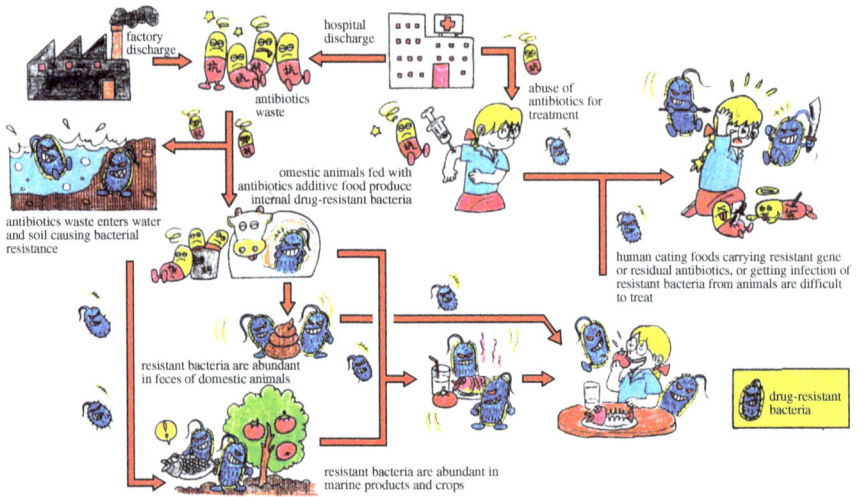

Figure 5-12: Biological chain shows residual antibiotics leading to development of bacterial resistance.

One, studies have found that abuse of antibiotics can induce some "silent" resistant genes. Under this circumstance, the originally covered feature becomes more apparent. A typical example is the mechanism of bacteria resistant to erythromycin. One of the most important mechanisms of the bacteria resistant to erythromycin is that the bacteria can produce the erythromycin methylase, which can transport a chemical group ($-CH_3$) to erythromycin target site on the ribosome to deactivate the ribosome-binding activity of erythromycin. However, in the absence of erythromycin, erythromycin methylase is a "silent" gene and this transporter does not exist. In the presence of erythromycin, this silent gene will be transcribed and translated. A considerable number of erythromycin methylase transporters are created in bacteria, as shown in Fig. 5-13.

Two, scientists have found that another mechanism of antibiotics inducing bacterial resistance is the bacteria that can produce SOS response after induction. The SOS response is the error-prone repair mechanism in bacteria. When antibiotics are applied to bacterial cells, normal cell metabolism and synthesis are inhibited or went wrong. However, if SOS response is activated, it is possible to fix these errors and reduce the mortality rate of bacteria so that the bacteria

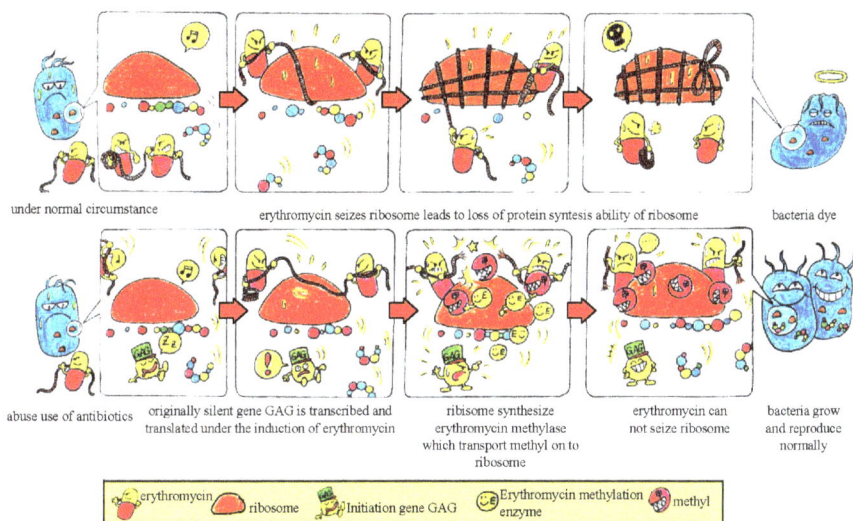

Figure 5-13: Induction of silent methylase gene resistant to erythromycin.

can grow and reproduce normally under the pressure of antibiotics. Certain antibiotics such as mitomycin C and quinolone can induce the SOS response in *E. coli.*

Three, scientists recently discovered a very important mechanism of antibiotics inducing bacterial resistance. They found that antibiotics can induce bacteria to become competent. Competent cells are cells that can easily absorb exogenous materials. In one experiment, streptomycin resistant gene fragment was added into the culture of *Streptococcus pneumoniae* in the presence of certain dose of streptomycin (625 ng/mL, this concentration should be precise to avoid the death of most of the cells). Same culture without streptomycin is used as control. The result showed that resistant bacteria emerged in the culture with streptomycin but not in the culture without streptomycin. This is the evidence that streptomycin can induce the cells to become competent and absorb the exogenous resistant DNA fragment to integrate into the sensitive bacteria and then make it resistant. Mitomycin C can also induce *Streptococcus pneumoniae* to become competent. Normal cells become competent cells to promote some transformation activities inside the cell.

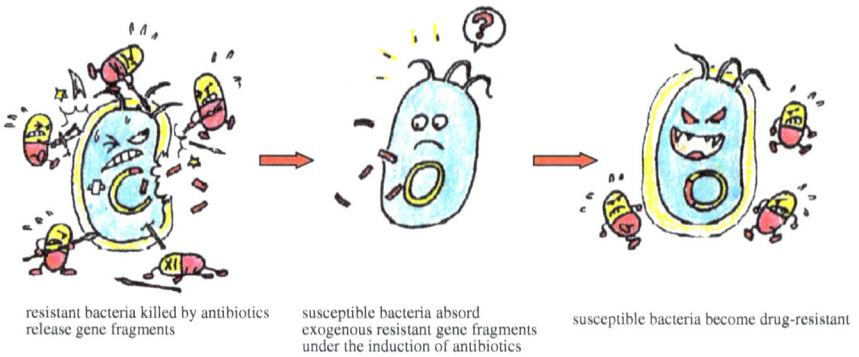

resistant bacteria killed by antibiotics release gene fragments   susceptible bacteria absord exogenous resistant gene fragments under the induction of antibiotics   susceptible bacteria become drug-resistant

Figure 5-14:   DNA fragment carrying resistant gene is absorbed by competent cells.

This scientific finding suggested that even those antibiotic-resistant bacteria were killed by enormous antibiotics, their death could release the resistant gene to the environment and be integrated by other sensitive but competent bacteria, becoming new resistant bacteria. Thus, the spread of bacterial resistance can be attributed to the abuse use of antibiotics by human, causing hazards everywhere. Considerable attention should be drawn towards antibiotic residues, resistant bacteria, and resistant genes that circulate in animals, plants, and the human body. Figure 5-14 shows that the resistant gene fragment is absorbed by competent bacteria.

In addition, latest studies by scientists found that most of the bacteria will be killed in the presence of antibiotics, but a considerably small number of bacteria can still survive. These resistant bacteria can produce a chemical substance called indole. This chemical substance can help sensitive bacteria to resist the activity of antibiotics, so that the entire population of bacteria becomes more resistant. This phenomenon is somewhat similar to a few power men (resistant bacteria) providing charitable donation to most of the weak men (susceptible bacteria). This causes resistance in the entire population of bacteria against antibiotics, as shown in Fig. 5-15.

"Substitution" causes "superbugs"

We know that in addition to their own reproduction and spreading of the resistant bacteria, another reason that causes "superbugs" is some

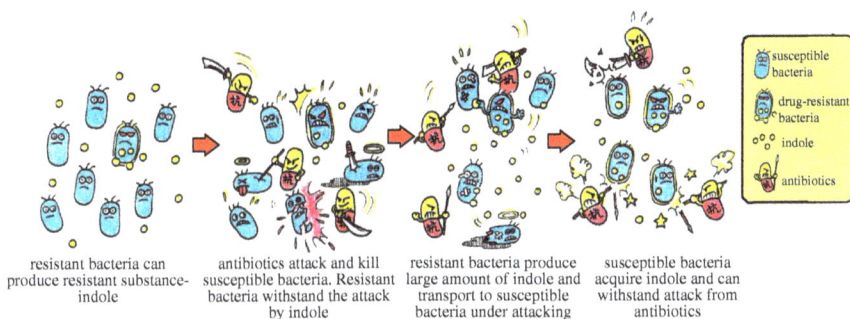

| resistant bacteria can produce resistant substance-indole | antibiotics attack and kill susceptible bacteria. Resistant bacteria withstand the attack by indole | resistant bacteria produce large amount of indole and transport to susceptible bacteria under attacking | susceptible bacteria acquire indole and can withstand attack from antibiotics |

Figure 5-15: Resistant bacteria help susceptible bacteria to defend antibiotics.

bacteria originally susceptible to antibiotics acquire resistant gene from resistant bacteria like "substitution" to promote their ability against antibiotics. Thus, what causes this "substitution"? Investigations have proved that the bacteria have plasmids, vectors carrying resistant gene, and transposon, genes capable of "dancing."

Plasmids are double-strand circular DNA capable of autonomous replication outside of the chromosome. They also carry genetic information and control certain characteristics, which are not essential for bacterial survival. Bacterial plasmid can self-replicate, that is, can be passed to the offspring. Several plasmids can coexist in a bacteria cell but also can naturally be lost. Another important feature of plasmid is that it is similar to a vehicle; it can bring resistance genes to other antibiotic-sensitive bacteria and let it continue to multiply and spread among bacteria. R plasmid has been found in bacteria that carry resistant gene, which controls the production of enzyme to inactivate drugs, or reduce membrane permeability to the drugs. Figure 5-16 shows the process of susceptible bacteria that acquire drug resistance through the transfer of plasmid carrying resistant gene.

Through the above process, *Staphylococcus aureus*, originally sensitive to antibiotic, acquires the plasmid carrying resistant gene from *Enterococcus faecalis* and becomes antibiotic resistant.

A transposon is a "jumping gene." This type of "dancing gene" was first discovered by Barbara McClintock (1902–1992, Fig. 5-17) at Cornell University. McClintock found a considerably strange genetic behavior when studying corn. One mutant gene can change the color

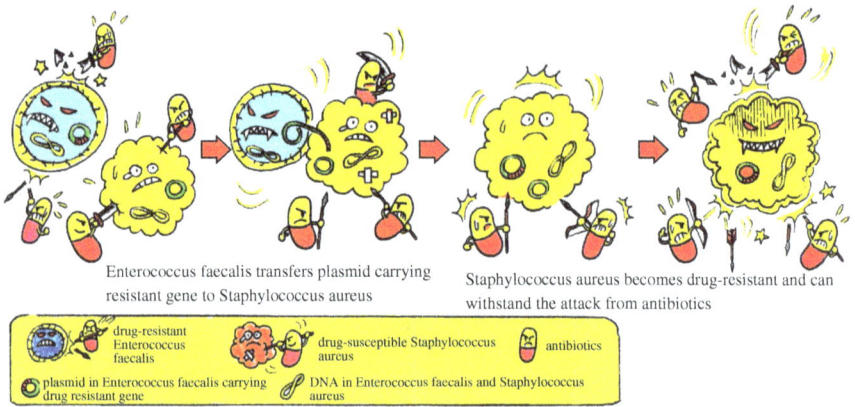

Enterococcus faecalis transfers plasmid carrying resistant gene to Staphylococcus aureus

Staphylococcus aureus becomes drug-resistant and can withstand the attack from antibiotics

drug-resistant Enterococcus faecalis

drug-susceptible Staphylococcus aureus

antibiotics

plasmid in Enterococcus faecalis carrying drug resistant gene

DNA in Enterococcus faecalis and Staphylococcus aureus

Figure 5-16: Susceptible bacteria acquire drug resistance through transfer of plasmid carrying resistant gene.

Figure 5-17: Barbara McClintock and "jumping gene" causing different corn colors.

of the corn. Moreover, it seems as if it can also jump from one cell to another one. When the cell obtains this gene, the other one loses it (Fig. 5-17). From 1944, she began to study this strange genetic behavior. It took her 6 years to engage further observation and experiment. In 1951, she presented her results in a symposium held at Cold Spring Harbor, US. She was expecting the genetics community to accept the research results and maintain positive attitude on it (because this is a big breakthrough in genetics, previously geneticists believed that the genetic composition of the organism is constant except the gene recombination that occurs during vertical transmission). But, unfortunately, she did not see the expectation, instead of mockery and neglect. Her great contribution was not accepted until 1970s when the development of novel molecular biology techniques revealed that scientists can isolate and characterize this type of mobility gene from different organisms. Thus, in 1983, she won the Nobel Prize in Physiology or Medicine at the age of 81.

McClintock was born in Hartford, Connecticut in the United States on June 16, 1902. Under the support from her father, she entered the School of Agriculture at Cornell University after World War I. She earned her bachelor's degree in 1923 and PhD in botany in 1927. For women of that era, she is unique at the frontier because she stayed in the field of biological research and continuously engaged herself in scientific research. But, even at the time, her decision was epoch-making, not in terms of the prevailing circumstances. She has never obtained a permanent position at the university. At the most desperate time in her life, she considered to give up her work and become a meteorologist. After obtaining her PhD degree at Cornell University, she served as a lecturer for 4 years and engaged herself in research in several institutes or universities since 1931. Until 1941, she obtained her first permanent job at the age of 39, became a research scientist in genetics lab at the Carnegie Institution of Washington. She worked there until the end of her life in 1992.

At present, a considerable number of transposons (Tn) carrying drug-resistant gene have been found in resistant bacteria. It has also been confirmed that Tn can easily transpose from the bacterial chromosome to some of the carrier (such as a plasmid). Therefore, Tn can quickly spread to other bacteria, which is one of the important causes for the emergence of bacterial resistance in nature.

In summary, the first main reason for the cause of drug-resistant bacteria and their spreading is that, after killing a large number of antibiotic-sensitive bacteria, a considerably small amount of resistant bacteria gain the possibility to grow without limitation. Second cause is that the use of antibiotics induces some silent resistant genes to make the sensitive bacteria become resistant. The third one is that the sensitive bacteria acquire vehicles (plasmids or transposons) carrying drug-resistant gene or directly absorb DNA fragment carrying drug-resistant gene to gain drug resistance and convert the sensitive bacteria to resistant bacteria.

Now, we have learned the procedure of bacteria becoming drug resistant. Next, we are going to illustrate the strategies and tactics used by resistant bacteria to defend themselves from the attack of antibiotics.

Strategies and tactics one — to change the material for building bacterial cell wall

One of the material bacteria used for building its cell wall is two linked alanine, stands for D-Ala-D-Ala. The "elite antibiotic," vancomycin, used in clinic to treat resistant bacteria now uses five "ropes" (hydrogen bond between the molecules) to bind this raw material for bacteria cell wall, and destroy the construction of the cell wall to achieve the effect of anti-infection.

However, those vancomycin-resistant bacteria can use the material that is different from the material used for building sensitive bacteria cell wall under the regulation of resistant gene. They can change one of the alanine to lactic acid (D-Ala-Lac), or to serine, or delete one alanine so that one of the five ropes vancomycin used for binding the cell wall material loses it activity. Thus, the materials are not effectively bound by vancomycin, and bacteria can use these materials for building the cell wall and run normal growth and reproduction. Figure 5-18 shows that the drug-resistant bacteria change their cell wall material so that vancomycin cannot properly bind cell wall and lose activity.

Strategies and tactics two — develop new cell wall "architect"

During the construction of cell wall, PBP (PBP consists of two "architects": transpeptidase and transglucosidase) is an important cell wall "architect". It is in charge of building the net structure of bacterial cell. In general, penicillin or similar drugs can accurately capture these "cell wall architect" then destroy the "wall" structure for the purpose of killing the bacteria.

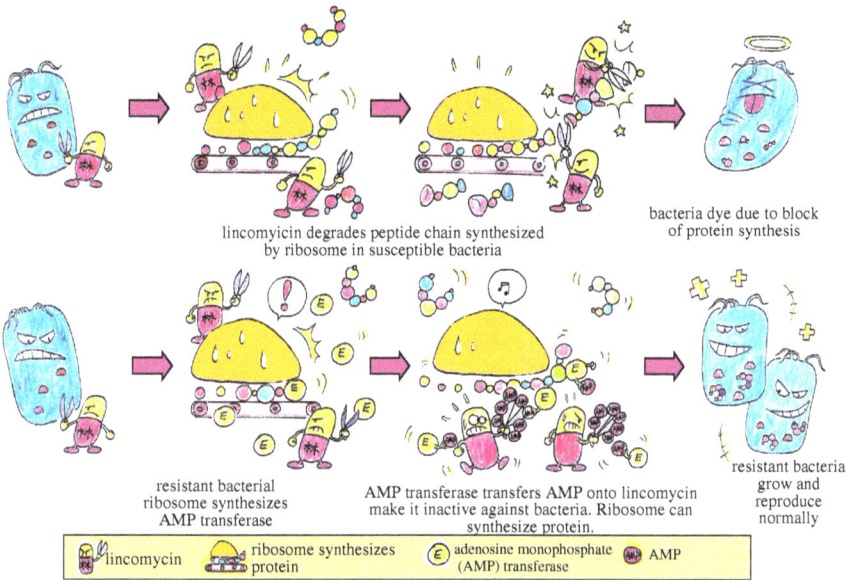

Figure 5-20: Mechanism of lincomycin losing its antibacterial activity.

lincomyicin degrades peptide chain synthesized by ribosome in susceptible bacteria

bacteria dye due to block of protein synthesis

resistant bacterial ribosome synthesizes AMP transferase

AMP transferase transfers AMP onto lincomycin make it inactive against bacteria. Ribosome can synthesize protein.

resistant bacteria grow and reproduce normally

lincomycin

ribosome synthesizes protein

adenosine monophosphate (AMP) transferase

AMP

Figure 5-21: Penicillin attacks susceptible and resistant bacteria.

penicillin attacks susceptible bacteria

penicillin seizes PBP

susceptible bacteria cell wall disintegrates and dye

penicillin attacks resistant bacteria

resistant bacterial ribosome synthesizes penicillinase to attack penicillin. Penicillin is degraded and lose antibacterial activity

resistant bacteria grow and reproduce normally

penicillin

PBP participates synthesis of bacterial cell wall

intracellular contents

ribosome

penicillinase

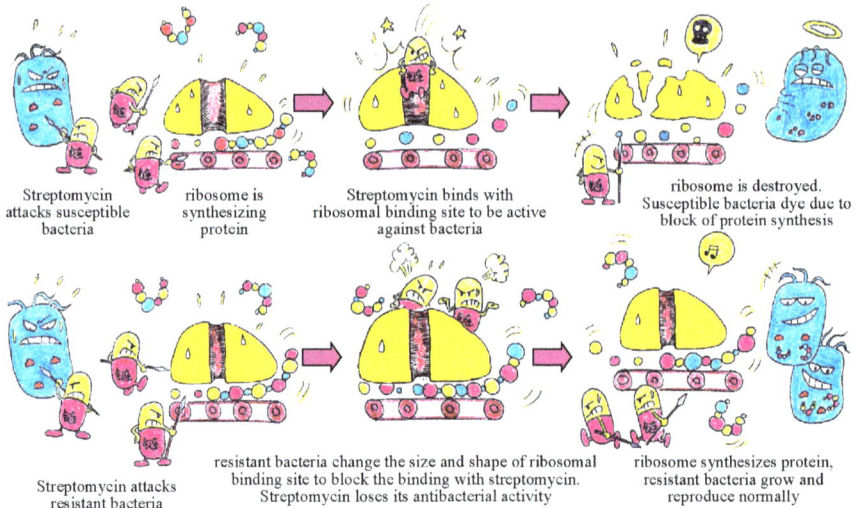

Figure 5-22: Bacteria change streptomycin targeting site to become resistant.

like the organs of human. If they were attached or destroyed the important life activities of bacteria will be blocked; thus, bacterial infection is eliminated. The protein factory, ribosome, of sensitive bacteria has a target site that can be attacked by streptomycin. When streptomycin reaches this site, this protein factory is destroyed and bacteria will be dead. However, this site in protein factory, ribosome, of resistant bacteria has changed so that antibiotics cannot reach the target and destroy the protein factory. Bacteria are still alive with regular growth and reproduction, as shown in Fig. 5-22.

Strategies and tactics five — strengthen the defense of main path; reduce the bullet entering the cell

Cell outer membrane is the first line defense of some Gram-negative bacteria such as *Pseudomonas aeruginosa* and *Mycobacterium tuberculosis.* It wraps up the outside of the bacterial cell wall and is a highly selective permeability barrier. There are some special proteins called porins on the cell outer membrane. They act similar to gates, allowing some molecules outside the cell to enter into the cell from

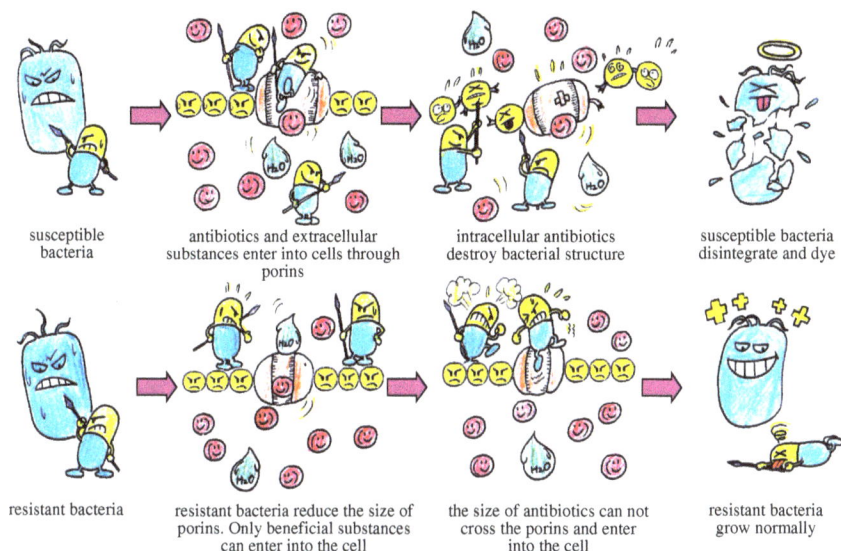

susceptible bacteria

antibiotics and extracellular substances enter into cells through porins

intracellular antibiotics destroy bacterial structure

susceptible bacteria disintegrate and dye

resistant bacteria

resistant bacteria reduce the size of porins. Only beneficial substances can enter into the cell

the size of antibiotics can not cross the porins and enter into the cell

resistant bacteria grow normally

Figure 5-23: Bacteria change porins to become resistant.

this channel. Normal bacterial cell outer membrane also allows antibiotics entering into the cell through the porins that play the role of efficacy inside the cell. However, in some drug-resistant bacteria, the number of porins is fewer or the channel becomes smaller or even shut off. This blocks or reduces the number of "bullet" — antibiotics entering into the cell, resulting in drug resistance, as shown in Fig. 5-23.

Strategies and tactics six — Manufacture efflux pump to bring bullets outside of the cell

Currently, some drug-resistant bacteria in clinics are found to have the "secret weapon," efflux pump, on their membrane. These efflux pumps can bring various "bullets" that has already entered the cell to the outside of the cell. Thus, antibiotics constantly enter into the cell. On the other hand, they are captured by grab-machine-like efflux pump and transported to the outside. Figure 5-24 shows drug-resistant bacteria transporting antibiotics outside of the cell by efflux pumps powered by adenosine triphosphate (ATP).

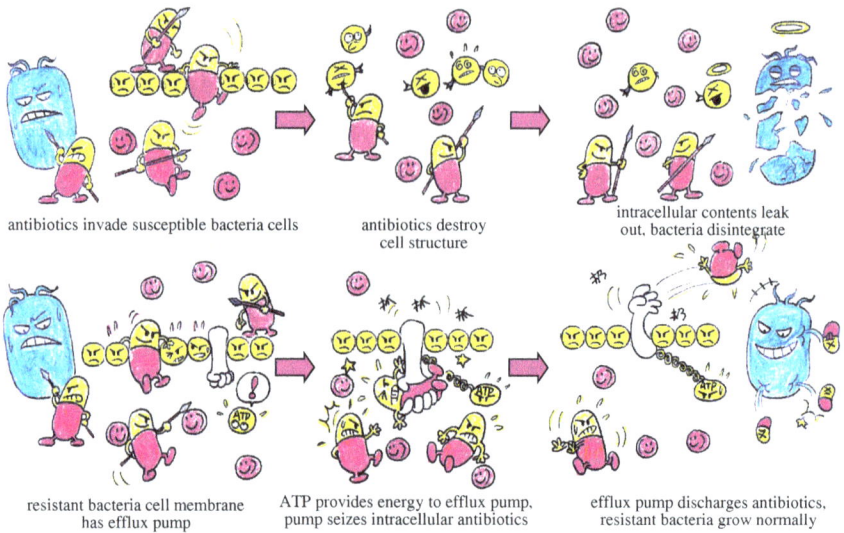

antibiotics invade susceptible bacteria cells   antibiotics destroy cell structure   intracellular contents leak out, bacteria disintegrate

resistant bacteria cell membrane has efflux pump   ATP provides energy to efflux pump, pump seizes intracellular antibiotics   efflux pump discharges antibiotics, resistant bacteria grow normally

Figure 5-24:   Bacterial "efflux pump" and drug resistance.

Strategies and tactics seven — gather together to produce invulnerable bullet-proof jacket

From late 1993 to early 1994, several hundreds of asthma patients suffered a mysterious bacterial infection in the United States. Any antimicrobial drug has no effect. Finally, at least 100 were dead, among which more than 50 family members of victims filed a lawsuit to the court. This incidence shocked the nation and led to a comprehensive investigation. Results showed that all patients had used ordinary albuterol inhaler. Therefore, people speculated that the device for drug production was contaminated with these mysterious bacteria during the process. However, the manufacturer believed that their production process was sterilized with chemical disinfectants, strictly in accordance with the related provisions. The final survey revealed that the container for drug production was contaminated with special bacteria, *Pseudomonas aeruginosa*, which not only can cause pneumonia, but also can secrete mucus gathering together to form mycoderma. Bacteria with mycoderma are resistant to chemical disinfectants, antimicrobial agents, and immune systems.

Bacterial mycoderma, a type of group colonization, is formed during growth to adapt to the living environment and are adsorbed onto the inert or active material surface. It is formulated by bacteria itself and extracellular matrix. Studies have shown that this type of bacteria group has strong resistance against antibiotics. They can also escape from the host immune system, and the infection is very hard to treat. It is one of the causes for clinical obstinate infection.

Formation of mycoderma and release of bacteria can be classified into five stages: the first stage is that bacteria find a suitable adhesion material and adhesion starts from single cell; the second stage is that grouped colonies begin to secrete extracellular substances, and further attract more bacterial adhesion; the third stage is that more bacteria group together and wrap up all the bacteria with the extracellular substance secreted by themselves. The bacteria inside start to be dormant; the fourth stage is that bacteria continuously produce a considerable number of extracellular substances and wrap more bacteria; the fifth stage is that those dormant bacteria inside revive and come out of the mycoderma in the presence of certain internal or external circumstances. Figure 5-25 shows the process of bacterial resistance caused by bacterial mycoderma.

antibiotics attack pathogens, pathogens adhere with each other under the pressure

extracellular substances secreted by pathogens form velum to defend attack from antibiotics

when conditions are improved, pathogens get out of velum and become pathogenic

pathogens are dormant in velum, they are not pathogenic

Figure 5-25: Bacterial velum and drug resistance.

The bacterial mycoderma appears similar to a solid bullet-proof jacket. Typically, antibiotics are difficult to pass through. On the other hand, most antibiotics are only active against the growing bacteria but not dormant bacteria. Thus, bacteria hiding in the mycoderma can survive for a long period of time. When these dormant bacteria revive and come out of mycoderma, they are still capable of attacking the human body. Bacteria capable of forming mycoderma is more dangerous to human, because it can be long-term "latent" and attack us surprisingly whenever we are in bad health or do not pay attention. Many clinically used implants, such as the various stents of organs, catheters, prostheses, and tissues are the best materials for bacterial adhesion. Once mycoderma is formed, it is considerably difficult to treat. For this reason, some implants should be removed periodically. Formation mycoderma is one of the causes for clinically delayed healing of chronic infection and failure of antibiotics treatment.

Chapter Six

Who will be the Winner?
The War Continues

After the invasion of human pathogens, whether it will lead to human disease or recovery from the disease depends on three factors, the pathogen, the human body, and the use of antimicrobial agents, as shown in Fig. 6-1. Pathogens undoubtedly play an important role in the disease onset, but it cannot decide the

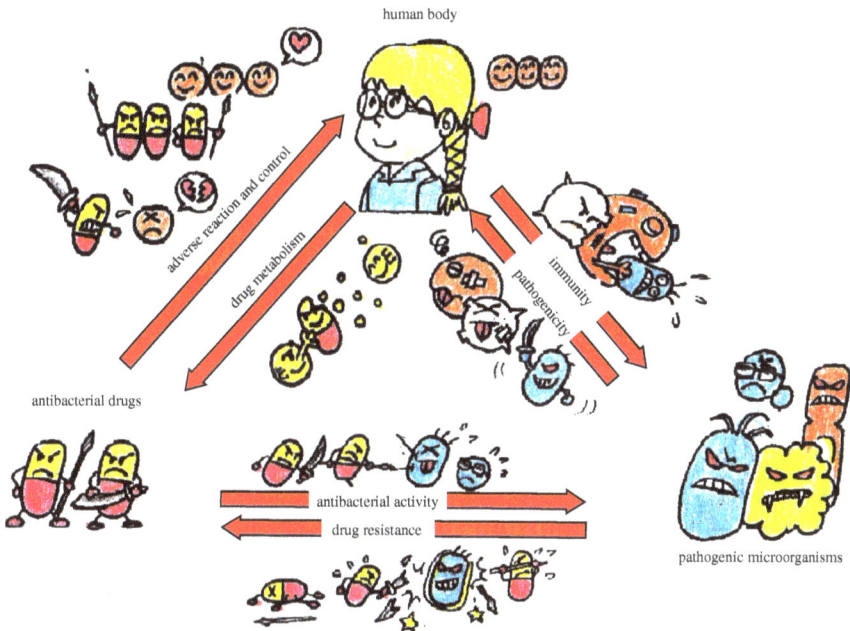

Figure 6-1: Interactions among human body, bacteria, and antibacterial drugs.

whole process. Human body's reactivity, immune status, and defenses are also important to the occurrence, development, and prognosis of the disease. When the body's defense system is predominant, we can defeat the pathogen and the pathogen will not cause the disease, or we can quickly recover from the disease. Bacteriostatic or bactericidal effect of antimicrobial drugs is the external factor to stop the progression of the disease and promote the rehabilitation. It is in favor of complete elimination of pathogens and recovery. Things always have two sides; contradictions are continually transformed. Under certain conditions, the bacteria can develop drug resistance, so the drug will lose antibacterial activity. A drug has the primary effect in a treatment, but it may also affect the health of the patients or cause treatment failure if not properly used.

During the long journey of conquering bacterial infection in human life, we have developed strategies from the empirical "Stone Age" to the rational "gain the win". This is a war between human and bacteria, which is full of "blood and tears". Winning or losing each battle is a challenge for human wisdom, and it is the result of the competition between human and cunning bacteria. The war between human and bacteria can be traced back to the "Stone Age". However, the discovery and application of penicillin started the "modern war" between humans and devastating bacteria. In this war, human beings celebrate the victory of the battle for the invention or discovery of a new weapon from time to time. However, they also feel depressed when bacteria develop new weapons and make them invulnerable again and again. Bacterial resistance to antimicrobial drugs has become a global problem. More and more types of resistant bacteria produce higher incidence of drug resistance, and the speed of resistance development makes the situation increasingly serious. The infections caused by resistant bacteria, so-called "superbugs", are big threats to human health. Bacterial resistance has become a hot topic worldwide in the 21st century.

Numerous facts have proved that the war between humans and bacteria will continue endlessly. To overcome this tense situation, we have only two ways to ultimately win the war. First is the rational use of drugs. Different weapon and different attacking methods should be used based on different enemies. Meanwhile, the application and extent of the nonclinical use of drugs should be strictly controlled. Second is the research and development of new weapons. We should incorporate cutting-edge scientific technologies to develop new drugs that can conquer "superbugs". In this manner, we can dominate the war and ultimately win. Otherwise, with so many drug-resistant bacteria, currently available antimicrobial drugs will lose their activities and humans will be forced to return to the "post-antimicrobial drugs era", the same as the dark ages before the discovery of antimicrobial drugs.

Therefore, what can we do to get to these two ways?

Strict Surveillance to Know the Enemies Promptly

Action projects in the United States

In 1999, under the lead of Health and Human Service (HHS) of the United States, a special workgroup was established by 10 federal bureaus and ministries to deal with drug resistance. This workgroup was moderated by US Centers for Disease Control and Prevention (CDC), Food and Drug Administration (FDA), and National Institutes of Health (NIH). This workgroup announced that the United States should take the responsibility and that the situation of public health can be improved through the following measures:

1. Responding to the threat from the drug-resistant bacteria promptly and effectively;
2. Simplfying and encouraging the development and appropriate use of problem-solving products;
3. Helping consumers and health workers to obtain information on drug resistance and the principles for appropriate usage, so that the products can be used safely and effectively. Furthermore, this practice can extend the life of the drugs;
4. Paying attention to and coordinating the scientific researches for the FDA to overcome the incidence of drug resistance.

The workgroup proposed a public health action plan to combat antimicrobial resistance in 2001. The success of this action plan depends on the cooperation of some entities such as state health agency and local government, universities, special societies, pharmaceutical companies, health care members, agricultural plants, and the public. We need to know the occurrence of resistant bacteria, the seriousness of the threat to public health, and the resistance trend along with time to promptly monitor the outbreak of drug-resistant bacteria and response feedback to clinic.

The surveillance report in China

During the past 7 years, China has conducted several surveillances. Although the methods used, the objectives, and the host units are different, all the reports showed us a common resistant bacteria distribution and trend map in China.

Staphylococcus

The major problem of *Staphylococcus* is methicillin resistance. Methicillin-resistant *Staphylococcus* (MRS) includes methicillin-resistant *Staphylococcus aureus* (MRSA) and methicillin-resistant coagulase-negative staphylococci (MRScoN). MRS can be observed in 80%–92% of the hospitalized patients. MRS is usually not sensitive to commonly used antimicrobial quinolones, aminoglycosides, and macrolides in clinic. MRS infection is difficult to treat, and all the β-lactam drugs are not effective. Not many drugs are available in clinic because the FDA has only approved vancomycin as a treatment agent for MRS.

Currently, glycopeptide drugs, vancomycin (once was called the elite antibiotic and the last defense) and teicoplanin, are the first-line drugs for MRSA treatment; Spinosad and linezolid were developed recently.

Enterococcus

Enterococcus has 18 species among which *Enterococcus faecalis* and *Enterococcus faecium* are related to human diseases. The predominant causes of human urinary tract infections, sepsis, endocarditis, purulent peritonitis infection, and trauma can be attributed to 80% of *Enterococcus faecalis* and 20% of *Enterococcus faecium*.

The predominant problems of *Enterococcus* are that they are highly resistant to vancomycin and aminoglycoside antibiotics. Vancomycin-resistant *Enterococcus* (VRE) and aminoglycosides-resistant *Enterococcus* (HLAR) are two resistant strains. The surveillance report of drug resistance from 26 hospitals in China showed that the VRE accounts for 0%–8% of all *Enterococcus* cases whereas HLAR accounts for 60%–80% of gentamicin-resistant strains. Drug resistance to *Enterococcus faecalis* is 2.95% and 0.83% to vancomycin and teicoplanin, respectively, and to *Enterococcus faecium* is 5% and 3%, respectively.

Escherichia coli and Klebsiella pneumoniae

Escherichia coli and *Klebsiella pneumoniae* are easy to produce extended-spectrum β-lactamase (ESBL). ESBL production is different in strains found in different areas and units. In 1994, ESBL productions in *Escherichia coli* and *Klebsiella pneumoniae* were 10% and 12%, respectively. In 2000, these numbers were 25% and 30%. In 2001, they increased to 35.3% and 32.7%. Penicillins, cephalosporins, and monoamides are usually not effective against ESBL-producing strains. These infections cause high mortality.

Recently, sensitivity of ciprofloxacin and other quinolones to *Escherichia coli* has decreased from 54% to 25%. This situation confirms that quinolone drugs cannot be used for the treatment of *Escherichia coli* infection. Of the 1448 *Klebsiella pneumoniae* strains, sensitivity rate to antibiotics other than ciprofloxacin is similar to that of *Escherichia coli*. With the extensive use of third-generation cephalosporins, detection of ESBL-producing strains is increasing every year. As ESBL resistance is mediated by plasmid transformation, transduction, conjugation, and so on, epidemic of this type of drug resistance can be controlled by limiting the use of third-generation cephalosporins to inhibit the production of ESBL.

Enterobacter cloacae

High yield of cephalosporinases is the main cause of multidrug resistance of *Enterobacter cloacae*. However, it was reported in the 1990s that *Enterobacter cloacae* can produce plasmid-mediated ESBL. Thus, the situation of drug-resistant *Enterobacter cloacae* has become more serious.

Nonfermenting Gram-negative Bacilli

In recent years, nonfermenting Gram-negative *Bacilli* infection has shown an upward trend in hospital, increasing from 41.2% to 47.9%. Among these, *Pseudomonas aeruginosa* infection has the most incidences in hospitals, accounting for 25.1% of total Gram-negative bacteria infection and 46.9% of nonfermenting Gram-negative *Bacilli* infection. *Acinetobacter* infection contributes to 31.0% whereas *Stenotrophomonas maltophilia* is 9.2%.

The sensitivity of *Pseudomonas aeruginosa* to 11 antibiotics is decreasing. Studies showed that from 1994 to 2001 its sensitivity to imipenem and ceftazidime decreased from 96% and 92% to 75% and 79%, respectively.

Resistance of *Acinetobacter* to commonly used antibiotics remains high; only imipenem and cefoperazone (Sulbactam) are still effective to more than 70% of *Acinetobacter* infections.

Stenotrophomonas maltophilia is resistant to most of the common antibiotics owing to various mechanisms of resistance. It is naturally resistant to imipenem because of the production of L1 metallo-β-lactamase. Surveillance results suggested that *Stenotrophomonas maltophilia* is mostly sensitive to ticarcillin/clavulanic acid, cefoperazone (Sulbactam), and ceftazidime.

Non-neglected potential threat: drug resistance of the subepidemic strains

In resistant surveillance, we should not only pay attention to the predominant pathogens causing infections and dangerous bacteria causing epidemic outbreaks, but also take actions to reduce or control the growth of their resistances. However, more attention needs to be paid to some subepidemic strains and potential drug resistance to eliminate the potential epidemic before the outbreak. Currently, dangerous strains such as vancomycin-resistant *Staphylococcus aureus* (VRSA), which is a subepidemic strain, should be given immediate attention. Although no reports have yet been published but effective preventive measures should be taken to strictly control the use of glycopeptide resistance–inducing antibiotics and to put an end to the emergence of VRSA in China.

Burkholderia, Flavobacterium, and *Alcaligenes* are widely found in nature and in hospitals. They can easily cause infection in immunodeficient patients. Most of these bacteria have several resistance mechanisms. It is hard to treat their infection in clinic, thus the rate of mortality is high.

Resistance surveillance can cover the deficit of late report of bacteria incidences. Using large amount of data gathered from resistance surveillance during a long period of time, regularity of major drug resistance in local area and unit can be summarized. Based on this information, rational treatment plan can be derived for decreasing the bacterial resistance. This is very important to effectively control infections in hospitals.

Rational Use of Powers to Block the Spreading of the Enemy

Global strategy on bacterial resistance by World Health Organization

Many pathogens that cause acute respiratory infection, diarrhea, measles, AIDS, malaria and tuberculosis, and other diseases are resistant to first-line drugs, with rates ranging from 0% to almost 100%. Some are even resistant to the second- and third-line drugs. In addition, hospital-acquired infections caused by drug-resistant bacteria, emerging drug-resistant viruses, and growing neglected drug-resistant parasites in people living in remote areas and under poverty have greatly increased the global burden. The problem of resistance to drugs, especially for drug-resistant bacteria, is a threat to global stability and national security.

At first, people believed that the development of a new antibacterial weapon can defeat pathogens. Since the beginning of 21st century, the confidence has gradually disappeared. The sources for new drug discovery have gradually dried up.

In 1998, the World Health Assembly (WHA) urged member states to take actions to encourage appropriate and cost-effective use of antimicrobial agents, including the measures to:

1. Prohibit the use of antimicrobial drugs without prescription from licensed physicians.
2. Improve practices to prevent the spread of infection and thereby the spread of resistant bacteria.
3. Strengthen legislation to prohibit the production, sale, and distribution of counterfeit antimicrobial drugs, and ban the sale of antibiotics on the black market.
4. Reduce the use of antibiotics in meat animals.

Since the launch of the initiative by WHA, many countries are increasingly concerned about the problem of bacterial resistance; some countries have implemented national plans of action to solve this problem. We cannot predict what will happen in the future, thus people are increasingly aware of the current action that should be taken to prevent future disasters in view of failure of the weapons. The question is "what to do" and "how to do".

Faced with this challenge, WHO global strategy for containment of antimicrobial resistance provides an intervention framework to delay the emergence of drug-resistant bacteria and decrease their spread. The main measures are as follows:

1. To reduce the spread of infection.
2. To optimize the way to get qualified antibacterial drugs.

   • To improve the use of antimicrobial drugs.

- To strengthen health systems and monitoring capabilities.
- To strengthen regulation implementation and legislation.
- To encourage the development of appropriate new drugs and vaccines.

This human-oriented strategy targets the group of people related with drug resistance and needs them to participate in solving the problem. These include physicians; pharmacists; veterinarians; consumers; decision-makers of hospitals, public health, and agriculture; professional societies; and the pharmaceutical industry. The key of action is to improve the use of antibiotics to block the spreading resistance.

Government health policies and health care systems should play an important role in the containment of antimicrobial resistance. Rational rules and regulations should be implemented and strengthened. Establishment of national cross-sectional authorities (including community health care workers, veterinarians, agriculturalists, pharmaceutical manufacturers, government, media representatives, consumers, and other interested parties) is necessary to raise awareness about antimicrobial resistance alert. The committee should organize data collection and oversee local executive staff. For practical purposes, there should be a government authority to collect data from various departments, to make sure that antimicrobial agents are only available through prescription, to establish mechanisms to facilitate physicians to execute the relevant rules and regulations, and to establish supervision system as well as to monitor the implementation of these rules and regulations.

In terms of development of drugs and vaccines, actions are required to encourage cooperation between industry, government, and research institutes to develop new drugs and vaccines and to develop optimized treatment while considering the safety and efficacy, and reducing the risk of drug-resistant bacteria screening. Meanwhile, the government should provide adequate patent protection of new antimicrobial drugs and vaccines based on intellectual property laws.

Rational use of powers to block the spreading of the enemy

Abuse of antibiotics not only will increase the incidence of adverse reactions and drug-induced disease, but also can cause patient's organ damage. This will not only destroy the body's normal flora, but also cause increase in bacterial resistance. If human beings continue to abuse the use of antibiotics, eventually there will be a day when certain diseases may become incurable owing to loss of efficacy of antimicrobial agents to bacterial resistance. Therefore, we should learn rational use of drug, stick to scientific use of drug, and prohibit abuse of antibiotics.

Rational use of antibiotics is a scientific question, so how can we use antibiotics rationally?

Because antibiotics can treat various infectious diseases, sometimes they are used as "panacea", no matter what diseases are treated with antibiotics. It is noteworthy that antimicrobial treatment is only effective against bacterial infections but not against viral infections. Most of the upper respiratory tract infection, sore throat, forcitis, and bronchitis are caused by viral infections; therefore, this type of disease should not be treated with antibiotics. If antimicrobial drugs are used for treatment, it is not only ineffective but also harmful because the chance of bacterial resistance is increased.

The effect of using antibiotics for every treatment is usually not ideal. The use of antibiotics should follow 3R (right) principles. The first R is the right time and the right opportunity; the second R is the right patient; the third R is the right anti-infection. Following the 3R principles, the cure rate of the patients can be increased. This can decrease the bacterial resistance in the environment and can also reduce the burden on patients and save social resources.

How to follow the 3R principles? The first step is to know the epidemic distribution of drug-resistant pathogens in the hospital and local areas. Each area and hospital needs to have an internal infectious pathogen management committee. This committee should disclose the local situation of drug resistance annually. Thus, the clinic physician can use the antibiotics more rationally. The second step is to know the situation of drug resistance due to major pathogens and the antibacterial activity of major antibiotics including the approximate half-life (time for half of the antibiotics to be metabolized in the body) of each antibiotic, which bacteria are sensitive to what antibiotics, and whether the antibiotics are time-dependent or concentration-dependent and their metabolic routes (via liver or kidney). This can maximize the activity of antibiotics after knowing all these pieces of information. The third step is to improve the extent classification of infectious diseases. For example, in division of respiratory disease, ventilator-associated pneumonia has light/moderate/severe classification according to the extent of infection. In division of transmitted infectious disease, there are many abdominal infections. Classification is also required for spontaneous bacterial peritonitis abdominal infections. These are labor-intensive work but require unified leadership of hospitals. In addition, proper antibiotics should be selected for patients according to their extent classification. The type of antibiotics used for treating has to be considered based on the patients' classification.

Governments should adopt restricted-use policies. Antimicrobial drugs can be used in rotation in hospitals. Some antibacterial agents can be discontinued after a period of time to restore their effectiveness against the bacteria.

Prophylactic use of antibiotics should be strictly controlled, and topical use of antibiotics (skin and mucosa) should also be avoided, because it will cause allergic reactions and easily cause the emergence of resistant bacteria, leading to poor efficacy for the treatment of systemic infections.

To control abuse of antimicrobial drug, what can you do?

First is to understand the concept that "abuse of antibiotics is harmful". Resisting the abuse of antibiotics is to defend your own health and life.

Second is to exercise, to enhance the body's immunity against bacterial infection; to develop good health habits, such as washing hands and eating well-cooked meals. Performing daily disinfection can also reduce the chance of infection and use of drugs. Antibiotics should not be used by yourselves even if you are sick. Do not ask physicians for prescription of antibiotics, and for "good medicine" or "overdose of medicine" as well. Always ask physician if it is necessary to use antibiotics. When using antibiotics, do not reduce the dose, stop the medicine optionally, or reduce the treatment time. When the effect is not satisfactory, dose and factors such as time and route of administration should be considered; do not arbitrarily replace new antibiotics.

Prevent from the starting point of food chain: reduce the use of antibiotics in animals

The problem of abuse of antibiotics has been controlled to a certain degree currently, but the overuse of antibiotics in animals cannot be ignored because this also hurts human health.

Animal feed is the first link in human food chain. Antibiotics have been used as feed additives in farm, which can not only improve the prevention and curing of diseases in animals, but also decrease the feeding amount and promotes the growth of the animals. However, this could result in resistance of bacteria in the animals. Thus, the antibiotics added in the feed will lose their activity against these resistant bacteria and this type of infection will lose control. After certain processing procedures, these animals will be put on the market and the remaining resistant bacteria may spread around through cross-contamination and infect human via food. Even if resistant bacteria have been killed through high temperatures before entering human's body, bacterial DNA or genetic materials can still invade into the body and spread the resistant gene to cause drug resistance. Alternatively, these resistant bacteria in animal may be transmitted to humans by contact. When humans are infected with these bacteria, drug resistance occurs when using antimicrobial drugs. This stretches the diffusion chain of bacterial resistance. Therefore, we should control the first link in food chain to make real effect in actions of stopping the abuse of antibiotics. This spread chain should be completely disrupted to overcome the threat to human health from the abuse of antibiotics. Figure 6-2 shows the spreading paths for resistant bacteria (epidemiology).

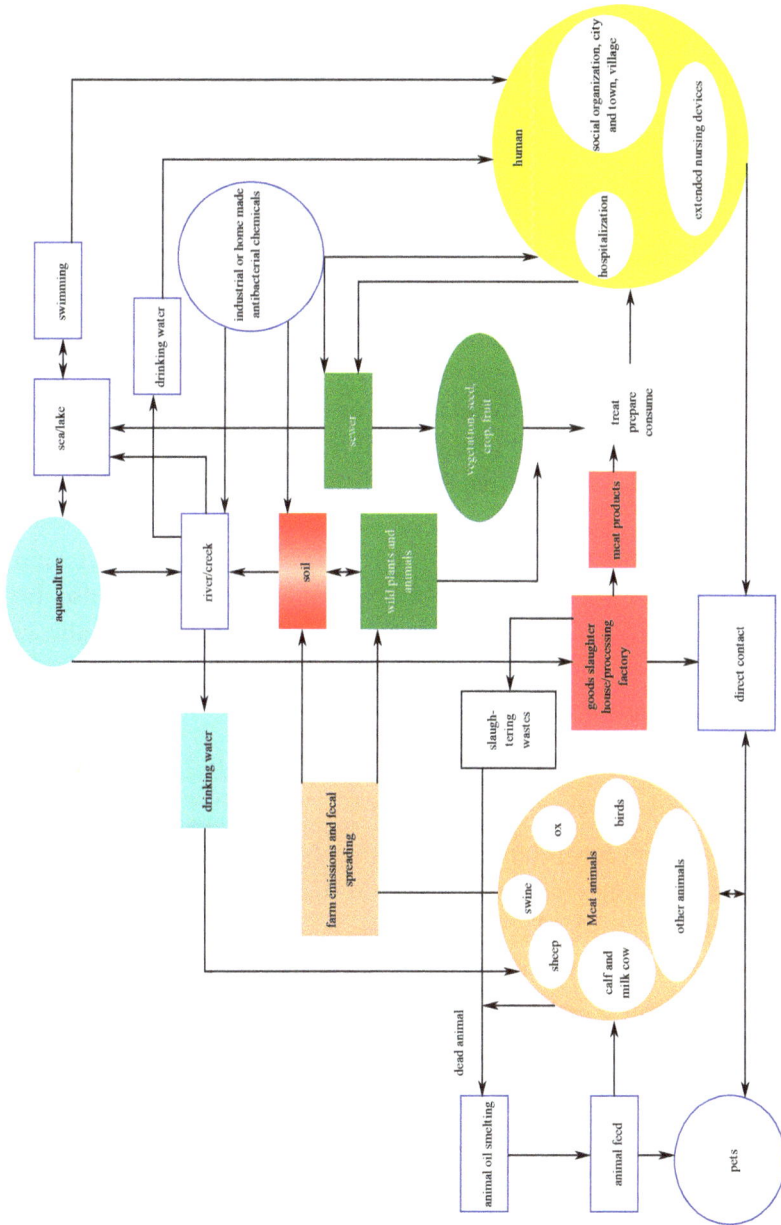

Figure 6-2: Spreading pathway of bacterial resistance.

Because we are still lacking comprehensive system of poultry industry and Drug Administration in China, supervising, monitoring authorities do not have legal documentation defining scope to control the antibiotics residues in the meats. Therefore, the use of antibiotics as feed additives has no supervision at all. The types of antibiotic used as feed additives almost include all antibiotics for human use. This directly makes *Escherichia coli, Staphylococcus,* and *Salmonella,* which are not typical pathogens in livestock, become major sources of infection in poultry.

Therefore, the government should set up monitoring and evaluation systems of food and drug safety and adverse reactions, and include the use of antibiotics in animals in this monitoring and evaluation system. Those antimicrobial agents that have been registered for human use can no longer be used as growth promoter for animals. Meanwhile, safe animal feed additives are encouraged to be developed.

Although the spread of resistant bacteria is transmitted between live bacteria, the resistance gene from dead bacteria can also be absorbed by live bacteria to cause resistance, and this should always be considered.

Continuous Discovery of New Weapons to Defeat the Enemy

As the battle between humans and bacteria becomes fierce, we are in the urgent need of new weapons to defeat the drug-resistant bacteria. How can scientists find effective weapons by using new technologies and resources to start new attack against the pathogens?

Strategy 1: expand microbial resources to search for new antibiotics

Screening of new antibiotics from microbes living in extreme environments

Extremophiles are extreme microbes living in extreme conditions such as low temperature, high alkali, high salt, and high pressure.

It is considered as one of the valuable sources for new antibiotics. Extremophiles are capable of producing a number of rare chemicals, from which new antibiotics with high activity may be discovered. Extremophiles include thermophilic microorganisms, psychrophilic, basophilic, eosinophilic, halophilic, and piezophilic microorganisms.

The optimum growth temperature of thermophilic microorganisms is higher than 45–50 °C, widely distributed in the hot springs, compost, soil geothermal areas, volcanic areas, and underwater volcanoes. Scientists even found a thermophilic microorganism in a hot spring in Iceland, which can grow at 98 °C. Psychrotrophic microorganisms are widely distributed in the polar regions of the earth, icehouse, snow mountains, deep sea, and permafrost regions. They can grow at 0 °C or even lower. Their optimum growth temperature is 15 °C or lower. In sulfur springs, sulfur mineral pyrite, or waste heap of gold, copper, lead, or uranium mine, the acidity is higher than homemade rice vinegar. Eosinophilic microorganisms live under such acidic conditions. Halophilic microorganisms are distributed in salterns, salt lakes, salted products, and the world famous Dead Sea. They can grow in the environment with 15%–20% of salt; some even can live in 32% salt water.

Screening of new antibiotics from marine microorganisms

Studies of antibiotics from marine microorganisms have a very long history, which can be traced back to the 19th century. In 1889, De Giaxa found that seawater can inhibit the growth of *Bacillus anthracis* and *Vibrio cholerae* and indicated that marine microorganisms may produce substances inhibiting bacterial growth. In the 1930s, scientists realized the inhibitory effect of the sea, and study of antibacterial substances from marine microorganisms has become popular.

The ocean nurtures life. There are a number of creatures in the ocean. After investigation of samples from several ocean research sites and analysis, an international group of marine biologists found that the number of microorganism types in the ocean may be 100 times more than previously estimated. The number could be tens of millions, which indicates that 20,000 microorganisms may be present in 1 L seawater. If a swimmer accidentally swallowed a mouthful of water, he will also swallow 1000 different types of microorganisms. Diversity of organism has suggested that the ocean is an important source for screening of new antibiotics. In fact, scientists have found many new compounds with antibacterial activity from marine microorganisms. They are different types of structure including macrolides, aminoglycosides, aminocyclitols, anthraquinone, alkaloids, and phenazine compounds. These compounds have activities against corresponding bacteria or fungi or cancer. Some of the mechanisms are different from those of the existing antibiotics, showing unique activities.

The sea is a blue fertile land and a treasure bank of precious biological resources. It has 80% of the entire species present on the earth. Marine microbes are the most abundant marine life resources. The sea provides the basis of a variety of new drugs, and new inspiration for development of antibiotics.

Hellio *et al.*[a] extracted several substances having antimicrobial activity from 16 species of marine microalgae from Antarctic waters, in which 9 substances could strongly inhibit the growth of marine fungi and Gram-positive bacteria. *Snow Dragon*, the scientific investigation ship from China Polar Institute, undertakes the important task of investigation of the polar regions. Various samples collected by *Snow Dragon* were studied in many Chinese research institutions. Scientists from Ocean University of China obtained one bacterium that has anti-tumor activity, eight bacteria that have antimicrobial activity, after screening of 101 psychrotrophic microorganisms from Arctic mud and seawater. There are also many other related reports. This indicates that extremophiles are a potential source of active substances and have broad prospects in basic research and development applications.

Figure 6-3 shows *Snow Dragon* used by researchers for the investigation of Antarctic, and the collector for the collection of marine

---

[a]Hellio C, Bremer G, Pons AM, Le Gal Y, Bourgougnon N. Inhibition of the development of microorganisms (bacteria and fungi) by extracts of marine algae from Brittany, France. *Appl Microbiol Biotechnol.* 2000, 54(4):543–9.

snow dragon scientific investigation ship

box type silt collector is collecting sedimentary samples from antarctic ocean surface

multi-channel gravity collector is collecting antarctic sedimentary samples out of the sea

antarctic ocean gravity collector is launching

columnar sample is coming out

collector is collecting sedimentary samples from antarctic lakes

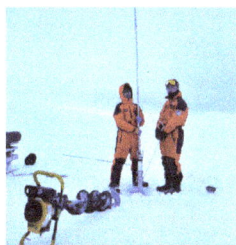

gravity collector is collecting sedimentary columnar samples under antarctic frozen lakes

Figure 6-3: *Snow Dragon* for south polar investigation in China and ocean silt collector (provided by Prof. Bo Chen).

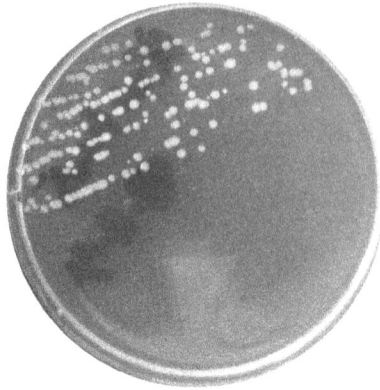

Figure 6-4: *Pseudomonas* in Antarctic seawater.

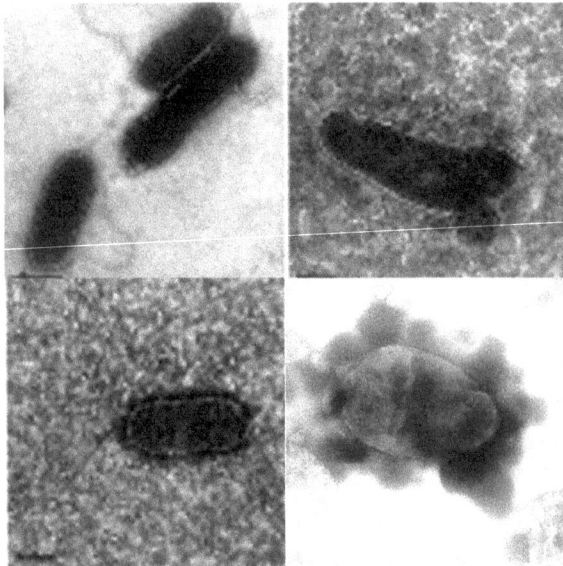

Figure 6-5: Electron microscopy of organisms from ocean sedimentary soil (provided by Prof. Bo Chen).

sludge. Figure 6-4 shows *Pseudomonas* in Antarctic seawater; Fig. 6-5 is some illustrations of electron microscopy of marine sedimentary soil microbial.

Screening of new antibiotics from plant endophytes

The emergence of drug-resistant bacteria urges the discovery of more and better antibiotics. At the same time, since the number of fungal infections caused by AIDS and organ transplantation is continuously growing, there is a high demand of effective antifungal agents. In addition, there is still a lack of drugs for protozoal and parasitic infection treatments. All of these situations require development of novel and effective drugs. In the past, the main source of antibiotics was soil microbes. Currently, plant endophytes have been considered as a great resource owing to their undeveloped diversity. Some studies of endophytes from plants that have pharmacological efficacy are important to determine whether these endophytes can produce substances with similar pharmacological efficacy as their host and to develop new culture technology and screening model for new pharmacologically active substances.

The so-called endophytes are fungi or bacteria that live in a variety of healthy plant tissues and organs at a certain stage of their life cycle or all stages of life. Infected host plants do not show (at least temporarily) disorders. Their endogenous living can be proved by isolation from strictly surface-disinfected histology or direct amplification of microbial DNA from plant tissues. They are normal endophytic bacteria living in the plant tissue and include not only mutually beneficial symbiotic microorganisms but also pathogenic microorganisms in the host plant. Almost all fungi isolated from the soil can be isolated from plants. Similar to the abovementioned *Actinomycetes* and bacteria, except for color changes, all other external shapes have no significant differences when cultured on solid medium (Fig. 6-6 shows the growth of fungi from soil on solid media); however, under an electron microscope, we can observe a variety of exotic fungal spores (Fig. 6-7 is electron microscopy of different fungi spores isolated from plant).

Endophytes live inside the tissues of plant. They have close contact with plants and can produce a variety of active biological substances, such as antibiotics. Thus, they are a promising resource for the discovery of new compounds, and endophytes have drawn attention as the new potential sources of therapeutics drugs or prodrugs.

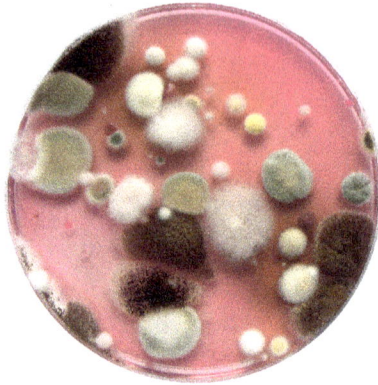

Figure 6-6: Soil fungi grown on solid medium.

In 1898, Vogl isolated the first endophytic fungi from the seed of ryegrass. However, in the next 70 years, the research progress of endophytes was slow, that is, until the 1970s when Bacon *et al.* found that endophytes in fescue are related to the production of toxins and the studies on endophytes became popular again.

Strategy 2: screening of new antibiotics from plants

Plant contains a number of medicinal resources. Although biological active substances currently found in plants are less effective than antimicrobial drugs used clinically, it is possible to find new chemical structure identities. For example, forsythia has broad-spectrum antimicrobial activity. Forsythol may be the major antibacterial ingredients. Puffballs acid extracted from puffball, also known as Mabo bacteria, has a broad spectrum of antibacterial activity against Gram-positive and Gram-negative bacteria, and fungi. Lonicera, also known as honeysuckle, strongly inhibits the secreted toxins from *Staphylococcus*, hemolytic *Streptococcus*, typhoid *Bacillus*, *Mycobacterium tuberculosis*, and *Pneumococcal*. Usnea, also called as tree hanging, cloudy grass, sea breeze rattan, canopy grass, and swallowwort, contains antibacterial substances. Its antibacterial spectrum is mainly Gram-positive bacteria and *Mycobacterium tuberculosis*. Usnic acid has strong activity against *Pneumococcus*, hemolytic *Streptococcus*, diphtheria *Bacillus*, and *Mycobacterium tuberculosis*. Blue cabbage leaf (Folium)

Figure 6-7: Electron microscopy of some plant endophytic fungi's spores.

contains tryptophan, isatin alkyl B, glucobrassicin, neoglucobrassicin, glucobrassicin-1-sulfonate, and indigotin. Leaves of *Strobilanthes* contain indigo glycosides, which then can be hydrolyzed to oxindole, and then be oxidized to indigotin. The whole grass of indigo plant contains indigo glycosides, yellow pigment, and tannin. They can inhibit the growth of *Staphylococcus aureus*, hemolytic *Streptococcus* to a certain extent. They are also effective against hepatitis B surface antigen and influenza viruses subtype A. In addition, indigo plant can also enhance phagocytic activity of white blood cells, reduce the permeability of skin capillary blood vessels, and has, among others, antiarthritic and antipyretic activities. Besides, the leaves of *Strobilanthes* can be used as drugs; its rhizome and root are used for preparing the well-known *Radix isatidis*. It has a significant detoxification effect. Geum plants contain indigo glycosides, Alexander anthocyanin, which is effective against *Shigella*, *Neisseria meningitidis*, and leptospirosis. European cranberry is famous as a "natural antibiotic". It can resist the attachment of *Escherichia coli* in the urethra and bladder wall, and the adhesion of *Helicobacter* as well. Bearberry leaf also has anti-*Escherichia coli* (urinary tract infection) activity. This has been discussed in Chap. 2.

Many herbs, and others including garlic and houttuynia, have such antibacterial activity. Although herbal antibacterial activity is relatively low, some antibiotic-resistant strains are still sensitive to a single herb. For example, scientists have found that traditional Chinese medicine dandelion, yellow cedar, and Patrina have various degrees of antibacterial activity *in vitro* against *Escherichia coli* and paracolobactrum, and particularly are still effective against multidrug-resistant bacteria. Doctors at Foshan First People's Hospital used eight types of Chinese herbal medicine to test their activity against MRSA and found myrobalan to be active against MRSA. Only a small amount of myrobalan can inhibit the growth of resistant *Staphylococcus aureus*.

Chinese herbal species are extremely rich in valuable resources. Development and utilization of Chinese herbs to find active substances that can inhibit the growth of bacteria or kill the bacteria is an effective way for treatment of drug-resistant bacteria.

Strategy 3: screening of new antibiotics from animals

Antimicrobial peptides are active against bacteria. They are produced by *Hyalophora cecropia* after Swedish scientist G. Boman *et al.*[b]

---

[b]Steiner H, Hultmark D, Engström A, Bennich H, Boman HG. Sequence and specificity of two antibacterial proteins involved in insect immunity. *Nature*. 1981, 292(5820):246–8.

injected *Enterobacter cloacae* and *Escherichia coli* in the worm to induce the production. These polypeptides are named cecropin. Since then, at least 1700 antimicrobial peptides have been found in insects, amphibians, aquatic animals, and mammals including humans, and in even plants and bacteria. They constitute the first line of defense for host against foreign pathogens infection.

Antimicrobial peptides have common characteristics that have a broad-spectrum activity (including Gram-positive bacteria, Gram-negative bacteria, fungi, and parasites) at low concentration, and activity against enveloped viruses, or even antitumor activity. Most antimicrobial peptides are nontoxic or less toxic to normal eukaryotic cells, and almost have no incidence of resistance. For this reason, antimicrobial peptides are also known as the "natural antibiotic". It is expected to overcome the growing problem of antibiotic resistance and is a hot research field.

Studies have found that most of the antimicrobial peptides are important components of natural defense system for animal against microbial infection. In humans and other mammals, these polypeptides like defensins are major proteins constituting neutrophils (approximately 10%–18%). Neutrophils are the most important cells to defend hosts against microbial attacks and acute infection. Other cells also can produce antimicrobial peptides.

At injured mucosal surface, including tongue, trachea, and intestines, a high concentration of polypeptides has been found. This may be an important mucosal protection component. Scientists found Maganins while studying why frogs and toads are free from infections in harsh environment. The production of Maganins and peptides from insects such as antimicrobial peptides is induced after injury. This induced response is found to be similar with the immune system of creatures.

Antimicrobial peptides have a unique antibacterial mechanism and are not prone to induced bacterial resistance. Recently, this has become a hot study topic. Researchers believed that cell membrane is the main target of antimicrobial peptides. Peptides can create pores in membranes by peptide–membrane lipid interaction. This destroys the structure of the membrane and causes

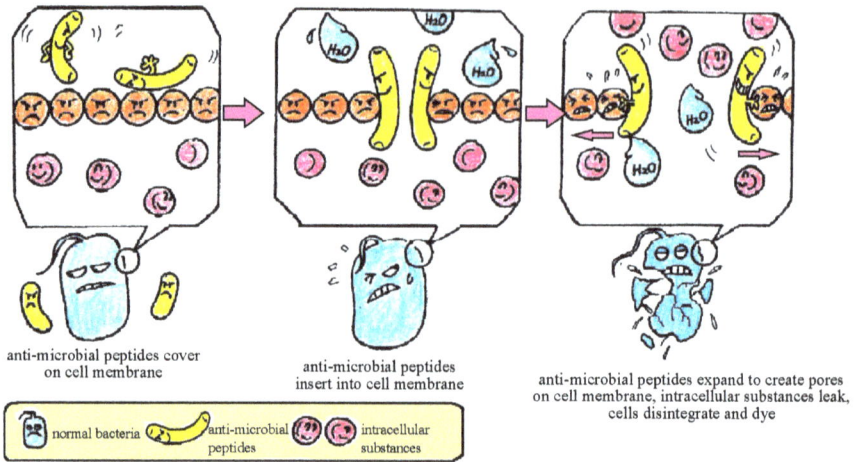

Figure 6-8: Antimicrobial peptides create pores on bacterial cell membrane.

imbalance between intracellular and extracellular osmotic pressures, leakage of cellular contents, and eventually cell death. The process is shown in Fig. 6-8, the antimicrobial peptides cover the cell membrane like a carpet. Then, single peptide molecule enters into the membrane. After several molecules enter the same site on the membrane, all peptides work together to create a "pore", a hole in the membrane.

Strategy 4: chemical synthesis of new chemical structure of antibacterial drugs

By chemical synthesis, scientists have developed a number of new structures of antibacterial drugs that are highly active against resistant bacteria. These include commonly used antibiotics, such as sulfanilamides, quinolones, and oxazolidinones, in clinics.

Sulfanilamides are synthetic antimicrobial drugs used for quite a long time and have been described in Chap. 4.

The most commonly used synthetic antibiotic in clinic is quinolones such as ofloxacin. Nalidixic acid is the first clinically used quinolone. It is not available owing to its narrow antimicrobial spectrum, poor oral absorption, and a number of adverse reactions.

Pipemidic acid has better activity against bacteria, better oral absorption, and less adverse reactions than nalidixic acid; it is available now for treatment of urinary tract infections and intestinal infections. Norfloxacin was synthesized in 1979, and a series of new fluorinated quinolones were synthesized subsequently, which opened a new chapter for quinolones. Quinolones can inhibit the activity of bacterial DNA gyrase and thereby block the synthesis of DNA leading to bacterial death. However, the bacterial resistance was found to occur with widespread use of quinolones. Studies confirmed that the main mechanism of resistance is mutations in chromosomes. There are two mechanisms of resistance: (1) change in bacterial DNA gyrase is related to high drug resistance of bacteria, and (2) change or deletion of the bacteria cell membrane porin channel is related to low drug resistance of bacteria.

In recent years, scientists have synthesized a new class of antibiotic — oxazolidinones. Oxazolidinones have a very broad-spectrum activity against Gram-positive bacteria. They are active against MRSA, VRSA, vancomycin-resistant *Enterococci*, penicillin-resistant *Pneumococci*, and anaerobes. The first launched drug in this class is linezolid (manufactured by Pfizer, marketed in 2000). It shows some clinical advantages in the treatment of Gram-positive bacterial infections, especially for the treatment of resistant bacteria.

Strategy 5: disarm the bacteria attacking the drugs

Currently, all clinical antibiotics (except for β-lactamase inhibitor) inhibit the growth and reproduction of bacteria. This "tit-for-tat" mechanism often results in "however persuasive good is, evil is still stronger". Even the activity of antibiotics against bacteria is more powerful; bacteria will try their best to produce defensive equipment to evade the attack by antibiotics. This leads to higher and higher possibility, and quicker and quicker emergence of bacterial resistance. Thus, an important and new strategy to combat drug-resistant bacteria is to disarm the weapons and equipment bacteria use to attack antibiotics. After these drug-resistant bacteria are disarmed, even commonly used antibiotics can kill the infectious bacteria. In addition, this type of new antibiotics will not kill bacteria, and normal bacteria will not develop resistance against them. The other strategy is to structurally modify the antibiotics so that they are not reachable to bacteria and their weapons and equipment.

Development of new antimicrobial drugs related to resistant-bacteria-produced hydrolase

Some drug-resistant bacteria can deactivate the antibiotics because they produce hydrolase to hydrolyze the drugs. If we can find substances that can inhibit bacteria-producing hydrolase, clinically used antibiotics may be active against those resistant bacteria again. As a matter of fact, some enzyme inhibitors have been used in clinic and achieved very good results. For example, Augmentin consists of amoxicillin and potassium clavulanate. Amoxicillin is a traditional β-lactam antibiotic to which a number of bacteria are resistant. However, potassium clavulanate is a powerful broad-spectrum β-lactamase inhibitor that can prevent β-lactamase destroying amoxicillin but without compromising its own antimicrobial activity. The combination of amoxicillin and potassium clavulanate produces good inhibitory effect on drug-resistant bacteria. There is another type of β-lactamase inhibitor that exerts nonreversible inhibitory activity against β-lactamase through competitive binding. It is active against β-lactamase produced by Gram-positive or -negative (except for *Pseudomonas aeruginosa*). It deactivates the enzyme by nonreversible reaction. β-lactamase is permanently deactivated even after the removal of the inhibitor. Under this circumstance, the inhibitor is also destroyed during the inhibitory process. Therefore, these inhibitors are also called suicide inhibitors. Moreover, because the activity becomes stronger along with the increased time, they are called progressive inhibitors as well. Sulbactam and clavulanic acid belong to this class.

Targeting at the deactivating enzymes produced by drug-resistant bacteria, scientists at NewBiotics have designed a new drug development program. They first identified a major drug-inactivated enzyme, and then modified the substrate of the enzyme to link with a toxic substance. When this substrate is hydrolyzed by the enzyme, it will release the toxic substance and cause bacterial death. Antimicrobial drugs developed using this strategy are called antimicrobial warriors.

It is also possible to modify the structure of existing antibiotics so that the bacteria weaponry (hydrolase) is unable to attach the antibiotics so that drugs are protected and are still active. Aiming at resistant bacteria producing β-lactamase, scientists have modified the existing β-lactam antibiotics to develop a series of penicillinase-resistant antibiotics. Penicillinase-producing resistant bacteria cannot attack these antibiotics.

Development of new antimicrobial drugs related to resistant-bacteria-produced antibiotics-inactivating enzymes

Aminoglycoside antibiotics are commonly used antibiotics. They have made an outstanding contribution in defeating certain bacteria. However, bacteria gradually become resistant to aminoglycosides. One of the most important resistant mechanisms is the use of some inactivating enzymes that can modify the structure of the aminoglycosides to deactivate the active ingredient of such antibiotics. Therefore, to reverse this situation, scientists have found two solutions. One is to introduce a protecting group in the structure of antibiotics so that the active site can be free from modification. In this direction, scientists have discovered amikacin. A protective group was introduced in kanamycin and the modified amikacin is no longer being attached by inactivating enzymes. Another solution is to delete the easy-to-be-attacked group in the structure. This is even better compared to introduction of a protective group because it is more difficult to lead to drug resistance later. Dibekacin is developed in this way and is a very effective anti-resistant-bacteria antibiotic.

Development of new antimicrobial drugs related to change of attacking site of resistant bacteria

To inhibit the growth of bacteria or kill the bacteria, antimicrobial drugs have to target the attacking site in bacteria. If the attacking site (target) is changed, drugs cannot effectively bind to the target and therefore they lose their activities. The strategy here is to increase the infinity between drugs and the changed target to effectively attenuate drug resistance. Carbapenems are a class of atypical β-lactam antibiotics. Their structures are similar to those of the traditional β-lactam antibiotics. However, the uniqueness of their structures can let them bind with changed/unchanged target, and they will not be hydrolyzed by β-lactamase.

Tetracycline is a broad-spectrum antibiotic active in inhibiting the growth of Gram-positive, Gram-negative, and other microbes. Currently, many bacteria are no longer sensitive to tetracycline. The reason is efflux mechanism and ribosomal protection of bacteria. The mechanism of action of tetracycline is binding ribosome to inhibit the synthesis of bacterial protein. Therefore, ribosomal protection leads to tetracycline resistance. If the synthesis of protein for ribosomal protection can be inhibited, tetracycline resistance can be managed accordingly. Scientists have modified the structure of tetracycline to develop glycyl-tetracycline (tigecycline). It can overcome the resistance owing to ribosomal protection.

Development of new antimicrobial drugs related to change of barrier in resistant bacteria

Bacteria have developed several defense mechanisms during combating with antimicrobial drugs. One of the mechanisms is related to the permeability of cell membrane (barrier). There are many porins in the outer membrane. These proteins can form specific or nonspecific channels on membrane that allow various substances including antibiotics to cross through the membrane. When the bacteria membrane loses certain porin proteins or changes the shape and amount of porins, the permeability of the membrane is also changed so that the infiltration of antibiotics is blocked. This mechanism of resistance is induced by altering the permeability of the membrane to create hindrance for drugs entering the intracellular compartment. This type of drug resistance can be overcome if the drugs can open up special channels in membranes to enter into the bacteria. Fortunately, scientists have invented a number of drugs following this idea. For example, imipenem, typical in the class, is an atypical β-lactam antibiotic. Its activity against *Pseudomonas aeruginosa* is achieved by drug infiltration from a special porin channel OprD. If bacteria can turn off this channel, then imipenem resistance may also occur. However, this idea is attractable and practical.

Development of new antimicrobial drugs related to drug efflux pump of resistant bacteria

One type of protein located on the bacteria cell membrane is called efflux pump. Researches have showed that this type of protein can pump out the antimicrobial drugs that have entered into the cell, just like a water pump does. Consequently, the concentration of intracellular antibiotics does not reach the limit to inhibit or kill the bacteria. This is one of the mechanisms how bacteria become resistant. To overcome this, scientists have found some efflux pump inhibitors to effectively inhibit the activity of efflux pump and ensure the intracellular concentration of antibiotics. One of the examples is tetracycline. Bacteria can very easily become resistant owing to efflux mechanism. The successor, tigecycline, can effectively overcome the resistance mediated by efflux pump.

Strategy 6: look for antimicrobial drugs with multi-killing mechanisms and new acting targets

Antibiotics with multi-killing mechanisms

Daptomycin is a new lipopeptid antibiotic isolated from the fermentation broth of *Streptomyces roseosporus*. It has a unique cyclic structure formed by linking a 10-carbon side chain and an N-terminal tryptophan of a cyclic β-amino acid peptide chain. Daptomycin exerts *in vitro* activity against all Gram-positive bacteria. This also includes all the resistant bacteria such as VRE, MRSA, glycopeptide-intermediate *Staphylococcus aureus* (GISA), coagulase-negative *Staphylococci*

(CNS), and penicillin-resistant *Streptococcus pneumoniae* (PRSP). Clinicians have very few choices of antibiotics to treat infections caused by these bacteria.

Daptomycin has a different inhibitory mechanism from other antibiotics such as β-lactams, aminoglycosides, glycopeptides, and macrolides. Daptomycin can destroy bacterial membrane functions in many aspects; however, it does not penetrate into the cytosol. Daptomycin-binding protein (DBPS) on membrane is the target site. Its possible mechanisms include inhibiting the synthesis of glycopeptides, inhibiting the synthesis of lipo-teichoic acid (LTA), and dissipating the electric potential of the membrane.

Looking for new drug targets

For the treatment of multidrug-resistant bacteria, finding new targets is an effective solution. Scientists have studied phospholipid synthesis pathway, fatty acid synthesis pathway, and peptide deformylase in order to develop new drug targets and to discover new antibiotics.

Strategy 7: discover new antibiotics by using genomics results

With the development of genomics, a new research model has emerged, that is, biochemical and functional studies are conducted on microbial genomic DNA sequences. Development of antibiotics also depends on this model. During the past 10 years, drug discovery has been more and more focused on the search for new targets-based technology. Potential chemical compounds are analyzed whether they are inhibitors of one certain biochemical reaction or molecular interaction (drug target) to test their drugability. This target-based model has obvious advantages. Information produced from genomics facilitates the selection of the drug target. Moreover, genomics-based new technology also brings new opportunities to find more effective antibiotics.

With the comprehensive study of the mechanism of action of antibiotics and resistance of bacteria, we gradually realized that one bacterium is a whole life. One life will utilize all protein networks to respond to external stimulation when the surrounding environment is changed, besides special response mechanism (target specificity). The changes in protein networks either cause cell death or enhance the survival of bacteria under the pressure of antibiotics, which leads to drug resistance afterwards.

Index

3R principles, 256

acquired immunity, 128
actinomycetes, 198
activated sludge, 62
acupuncture, 157
acute infection, 142
adverse effects of antibiotics, 207
agricultural antibiotics, 57
Alexander Fleming, 176
amino acid, 44
aminoglycoside antibiotics, 41
aminoglycosides, 204
aminoglycosides-resistant
    Enterococcus (HLAR), 251
anaphylactic shock, 208
antagonism, 41
anthracycline antitumor
    antibiotics, 42
anthrax spores, 81
anthrax vaccine, 164
antibacterial mechanism of
    penicillin, 219
antibiotic(s), 41, 196
antibody, 134
antigen, 137
antimicrobial warriors, 273
Antonie van Leeuwenhoek, 25
azithromycin, 205

β-lactam antibiotics, 187
bacillary dysentery, 152

Bacillus anthracis, 81
Bacillus thuringiensis, 55
Bacillus, 3
bacteria, 5
bacterial colony, 24
bacterial mycoderma, 245
bacterial vaccine, 168
Barbara McClintock, 235
biochemical pesticides, 57
biodiesel, 73
bioenergy, 69
bio-fertilizers, 58
biohydrogen production, 71
biological pesticide, 54
biological pigment, 49
bioremediation technology, 65
black death, 87
blood placental barrier, 131
blood–brain barrier, 130
Botulinum toxin, 107, 124
brewery, 53
Brucella, 109
bubonic plague, 80

Campylobacter jejuni, 106
capsular, 19
carrier status, 141
cell factories, 77
cell membrane, 21
cell wall, 19
central dogma, 217
cephalosporin C, 203

cholera toxin, 125
cholera vaccine, 162
cholera, 84
chronic infection, 142
Clostridium botulinum, 107
Clostridium perfringens, 104
Clostridium tetani, 105
coccus, 2
collagen, 121
collagenase, 121
competent cells, 233
complement, 134
copper utensils, 148
copper's antibacterial function, 150
Corynebacterium diphtheria, 95
cowpox, 160
cupron, 150
"curved-neck flask" experiment, 29
cytoplasmic inclusions, 22

dairy products, 53
dead bacterial vaccine, 168
diazotroph, 59
diphtheria, 95
diphtheria toxin, 125
DNA, 19
DNA replication, 221
DNA vaccine, 170

E. Jenner, 160
efflux pump, 243, 275
EHEC O104:H4, 113
endotoxin, 125
endotoxin shock, 127
Enterobacter cloacae, 115, 252
enzyme, 45
eosinophilic microorganisms, 261
eosinophils, 133

Erysipelothrix rhusiopathiae, 108
erythromycin, 205
ESBL, 251
Escherichia coli, 251
Escherichia coli variant O157, 112
ESKAPE, 115
ethanol fuel, 72
exotoxin, 123

fibrin clot, 121
flagellum, 19
Florey and Chain, 181
founding father of bacteriology: Pasteur, 27
Francisella tularensis, 109

genomics-based new technology, 276
Gerhard Johannes Paul Domagk, 171
Gram-negative, 20
Gram-positive bacteria, 20
Guasha therapy, 155

halophilic microorganisms, 261
hang wormwood for dragon boat festival, 151
healthy carriers, 93
Helicobacter pylori, 100
helicobacter, 4
hemolysin, 122
Hemolytic streptococcus, 97
human immune system, 127
hyaluronic acid, 120
hyaluronidase, 121
hydrocarbon, 50

immunoglobulin, 134
immunological prevention, 163
immunology, 33

infection, 141
inflammation, 134

jumping gene, 235

Klebsiella pneumoniae, 251
Koch Law, 116
Koch postulates, 36
koji, 53

Legionella, 111
Legionnaires' disease, 111
leprosy, 89
Lincomycin, 240
lipopolysaccharide, 21
lipoprotein, 21
Listeria, 108
live bacterial vaccine, 168
lysis, 218
lysosomes, 133
lysozyme, 177

macrolide antibiotics, 41
macrophages, 132
marine microorganisms, 261
marsh gas, 70
messenger RNA (mRNA), 217
metal β-lactamase antibiotics, 114
methicillin-resistant coagulase-
    negative staphylococci
    (MRScoN), 250
methicillin-resistant *Staphylococcus
    aureus* (MRSA), 228
microbial fuel cell, 74
microecological preparations, 47
monocytes, 132
Mrs. Montagu, 160
multi-drug-resistant bacteria, 229
mutualism, 41

Mycobacterium leprae, 90
Mycobacterium tuberculosis, 88

Nano silver, 147
natural immunity, 128
Neisseria gonorrhoeae, 99
Neisseria meningitidis, 98
neutrophils, 133
New Delhi-Metallo-1 (NDM-1), 114
nonfermenting Gram-negative
    Bacilli, 252
normal flora, 40
nucleotides, 45

obligate anaerobic bacteria, 105
opportunistic pathogenic bacteria,
    80
organic acid, 44
organic solvent, 45

passive immunity, 169
pasteurization, 32
pebrine
penicillin, 176
penicillin-binding protein (PBP),
    218
peptide vaccine, 170
peptidoglycan, 20
persistent lung disease, 152
pertussis, 168
petroleum exploration, 50
petroleum extraction, 50
phosphate-solubilizing bacteria, 59
phospholipids, 21
pili, 18
pioneer of bacteriology: Koch, 33
plant endophytes, 265
plasmids, 235
poisonous spinach, 113

polyhydroxyalkanoates (PHAs), 64
polyhydroxybutyrate (PHB), 64
polysaccharide vaccine, 171
porins, 242
potassium bacteria, 60
prebiotic, 48
principles of vaccination immunity,
  163
probiotics, 47
Prontosil, 171, 173
protein synthesis, 216
protein translation, 218
protein vaccine, 170
Pseudomonas aeruginosa, 100
pyemia, 143
pyogenic bacteria, 96

Qi Ai, 154
quinine, 155

rabies vaccine, 166
recombinant vaccine, 169
renewable energy, 69
resistant bacteria, 227
rhizobia, 59
ribosomal RNA (rRNA), 217
ribosome(s), 22, 217
rifamycin, 223
RNA transcription, 218

Salmonella, 103
Salmonella typhi, 93
sarcina, 3
seal fetus incident, 211
Selman A. Waksman, 189
semisynthetic antibiotics, 203
semisynthetic cephalosporin, 203
Shigella dysenteriae, 102
silent infection, 141

silver chopsticks, 145
smallpox, 159
solid surface culture, 185
SOS response, 232
specific immunity, 136
spontaneous generation, 29
spontaneous generation, 30
spore, 22
Staphylococcus aureus, 96
Staphylococcus epidermidis, 97
staphylococcus, 3
Streptococcus mutans, 98
Streptococcus pneumoniae, 98
Streptococcus suis, 110
streptodornase, 122
streptokinase, 121
streptomycin, 189
submerged fermentation, 185
suicide inhibitor, 272
sulfonamides, 171
superbugs, 114, 227, 248
synthetic bacteria, 77
synthetic biology, 77

tetanus toxin, 124
tetracycline antibiotics, 41
tetracycline teeth, 210
tetracyclines, 204
thalidomide, 211
the golden era of antibiotics, 201
toxic effects of antibiotics, 207
toxoid, 168
transcription, 216
transfer RNA (tRNA), 217
translation, 217
transposon, 235
tuberculosis, 87, 152
Typhoid Mary, 91
tyrothricin, 196

vaccine, 159
vancomycin-resistant
    Staphylococcus aureus (VRSA), 228
veteran disease, 111
Vibrio cholerae, 85
vinegar, 52
vitamin, 42

western medicines, 155
white plague, 87
white pollution, 64
World Tuberculosis Day, 89
wormwood, 153

www.ingramcontent.com/pod-product-compliance
Lightning Source LLC
Chambersburg PA
CBHW050545190326
41458CB00007B/1923